District Cooling Systems

This book draws on the authors' industry and academic expertise to explain the theory and practice of district cooling systems (DCS). The in-depth exploration of the design and development of DCS presents detailed best practices for their optimization in both the development and operation phases.

Readers will gain in-depth practical knowledge on all areas and considerations related to DCS technology's best practices, including current practical research areas and future potential research areas. This book addresses five areas related to DCS: the fundamentals of DCS technology, design optimization for development purposes, real-time optimization for daily operations, techno-commercial decision-making framework, and industry best practice. This information is presented through analyses of technological progress to date; case studies of current operations; and in-depth discussions of the theoretical bases and commercial, technical, and environmental benefits. Through this book, readers can recognize and apply best practices for the design, development, and operation of an optimal DCS design based on multiple factors, including financial analysis, energy efficiency considerations, and practical operation issues. This will enable them to contribute to national and international sustainable development goals regarding sustainable cities and climate action.

As this book provides both industry know-how and future research directions related to DCS, it is invaluable for DCS industry professionals and advanced undergraduate and postgraduate engineering students who aim to enter this industry and develop leading, highly efficient DCS systems. Overall, it is a vital resource for anyone involved in the planning, execution, and management of DCS projects.

District Cooling Systems

A Practical Approach and Application for Design, Operations, and Best Practice

Khin Zaw
Aung Myat
Md Raisul Islam
Poh Tiong Keng
Aung Kywe Nyunt

CRC Press
Taylor & Francis Group
Boca Raton London New York

CRC Press is an imprint of the
Taylor & Francis Group, an **informa** business

Designed cover image: @ Shutterstock

First edition published 2025
by CRC Press
2385 NW Executive Center Drive, Suite 320, Boca Raton FL 33431

and by CRC Press
4 Park Square, Milton Park, Abingdon, Oxon, OX14 4RN

CRC Press is an imprint of Taylor & Francis Group, LLC

ISBN: 978-1-032-67014-0 (hbk)
ISBN: 978-1-032-67015-7 (pbk)
ISBN: 978-1-032-67016-4 (ebk)

DOI: 10.1201/9781032670164

Typeset in Palatino
by SPi Technologies India Pvt Ltd (Straive)

Contents

Preface

The development of district cooling systems (DCS) is gaining continuous global attention due to their potential as sustainable cooling solutions that create commercial values; drive technical advancements; and support environmental, social, and governance (ESG) initiatives. This book aims to contribute to these goals by providing essential support to engineers and professionals in the DCS industry. It addresses two critical phases: (i) the project design phase, offering insights on both DCS equipment and optimal design strategies and (ii) the operation phase, delivering information on real-time operational optimization. The primary focus is on the design, development, and operation of DCS at optimal commercial and technical efficiency. Key components of DCS covered in this book include chillers based on vapor compression technology; the improvement of refrigerants, pumps, and cooling towers; and the design of chilled water distribution networks and machine learning optimization algorithms. These elements are fundamental to achieving excellent DCS system design and operations.

Currently, many DCS systems are in operation worldwide. However, there is a notable absence of detailed guidelines and best practices for the design and operation phases. A key distinction in the design of DCS, compared to conventional cooling systems, is the utilization of thermal energy storage. This technology is crucial for the effective and efficient use of capital expenditure (CAPEX) and for maintaining efficient operations throughout the entire economic life cycle of a DCS. Despite their importance, the benefits of DCS and thermal energy storage are not widely recognized within the industry. Furthermore, the principles of operational optimization are not well defined. This book aims to address these gaps by providing comprehensive information and guidance. It is important to note that DCS systems require more efforts during the design phase than conventional systems. This often necessitates technical design optimization along with financial analysis and economic life cycle savings analysis. The insights provided in this book will help bridge these gaps and support the development of more efficient and effective DCS systems.

The primary goals of this book are to equip engineers and professionals with comprehensive and practical information for the DCS design process, equipment selection, financial decision-making, and operational optimization. Additionally, it introduces machine learning, a type of artificial intelligence, as a critical tool for real-time operational optimization, which is essential for maximizing energy-saving potential in DCS operations. Properly designed and intelligently operated DCS systems not only enhance satisfaction but also uphold ESG principles, contributing positively to our environment.

Dr Khin Zaw
On Behalf of the Authors
22 Oct 2024

About the Authors

Khin Zaw is the Director and head of Energy-as-a-Service Technology the Infrastructure Division at Keppel and is a leading figure in Singapore's district cooling landscape. He was Associate Professor at the National University of Singapore (NUS) from 2019 to 2023. He holds a PhD from the NUS and a B. Eng degree from Yangon Technological University.

Aung Myat is an assistant professor at Tennessee State University. He obtained his Bachelor's degree from Yangon Technological University in 2004 and his master's and PhD from the National University of Singapore in 2006 and 2012, respectively. He is also a Singapore Certified Energy Manager and a Chartered Engineer with the Institution of Mechanical Engineers.

Md Raisul Islam is an associate professor at the National University of Singapore (NUS). He was awarded a research scholarship by the NUS in 1998 to undertake a research project on the heat and mass transfer processes of air-conditioning systems and was awarded a PhD degree. He is a chartered engineer.

Poh Tiong Keng is an executive director in the Infrastructure Division at Keppel. He is a registered professional engineer in Singapore. He holds an MBA (Dean's List Recipient) and a Bachelor of Engineering (2nd Upper, Hons) from Nanyang Technological University of Singapore and a Master of Science from the National University of Singapore.

Aung Kywe Nyunt is a senior associate Director at Beca Carter Hollings & Ferner (S.E. Asia), Pte Ltd. and has a three-decade career in engineering, procurement, and project management. He holds a Bachelor of Engineering (Mechanical) from the Yangon Institute of Technology and a Master of Science (Mechanical) from the National University of Singapore.

1

An Overview of District Cooling and Heating Systems

1.1 Introduction to District Cooling and Heating Systems

District cooling and heating systems (DCHS) emerge as pillars of innovation within the complex network of urban infrastructure, where the dual objectives of energy efficiency and sustainable solutions are met. This chapter thoroughly provides an overview of various aspects of DCHS, such as the key features of DCHS and the evolution of different generations of DCHS. Subsequently, this chapter highlights the advantages of DCHS. Lastly, the global market outlook of DCHS is further discussed. Therefore, this chapter provides the fundamental nature of DCHS, investigating its profound importance, complex mechanisms, and capacity for transformation.

1.1.1 Defining District Cooling and Heating Systems (DCHS)

DCHS represent the highest level of advancement in centralized air-conditioning technology. It is carefully designed to effectively manage the varied cooling and heating requirements of a cluster of buildings located inside a designated district. Fundamentally, DCHS operate by creating chilled water to facilitate cooling and hot water or steam to provide heating, so establishing itself as the thermal infrastructure that envelops the urban environment. The district is traversed by a sophisticated network of secondary water loops that precisely distribute thermal energy to specific buildings, therefore fulfilling their unique cooling and heating needs, as shown in Figure 1.1. District energy systems implement economies of scale through the consolidation of energy requirements from numerous buildings. This not only reduces energy costs but also enables the implementation of high-efficiency technologies like combined heat and power (CHP), which contribute to the overall improvement of energy efficiency.

DOI: 10.1201/9781032670164-1

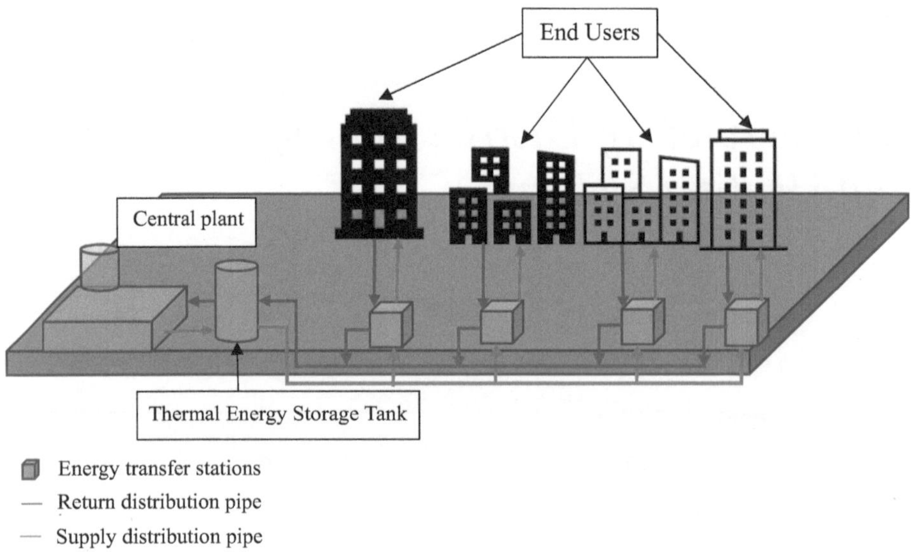

FIGURE 1.1
Schematic diagram of a district cooling and heating system (DCHS).

DCHS possess significant importance beyond its economic benefits, encompassing aspects of environmental stewardship and dependability. DCHS exemplify steadfast dependability by providing urban residents with uninterrupted thermal comfort, even when confronted with severe weather circumstances. The aforementioned dependability promotes the resilience of metropolitan areas and aids in the continuous operation of vital infrastructure. In addition, DCHS exemplify a profound dedication to ecological accountability. By utilizing DCHS, an alternative to the traditional method of installing individual chiller plants in each structure, a collective energy-saving solution is exemplified. This system not only mitigates environmental effects but also conserves resources and decreases greenhouse gas emissions. It improves the quality of life for city residents while bringing urban climate management into a greener, more sustainable era, serving as a beacon of sustainability.

1.1.2 Definition of District Heating Systems (DHS)

A DHS is a centralized system that delivers heating to a group of buildings or a whole district from a single or several heat sources. The production of hot water, steam, or other suitable heat carriers takes place in centralized plants through the utilization of a variety of heat sources. The entire DHS comprises three pillars, namely, (i) a centralized production plant, (ii) a distribution pipe network, and (iii) terminal stations or energy transfer stations. The centralized plant produces heat via multiple heat sources. These heat sources include CHP systems, geothermal energy, biomass or natural gas boilers, waste-to-energy plants, solar thermal collectors, and water heat recovery systems. The piping networks are insulated and serve as a backbone for circulating the heat carriers to specific terminal stations, also known as energy transfer stations, located near the buildings. The buildings have a dedicated water loop to carry the heat from the terminal stations to fulfill the heating needs, such as space heating, water heating, and other thermal purposes in the buildings.

1.1.2.1 Key Features of District Heating Systems (DHS)

A DHS encompasses several key features that distinguish them as efficient, sustainable, and reliable solutions for providing thermal energy to communities. Some of the key features include the following:

- Centralized Heat Production: DHS centralizes the production of heat in one or multiple centralized plants, typically located on the outskirts of urban areas or industrial zones. This centralized approach allows for efficient utilization of energy resources and facilitates the integration of diverse heat sources.
- Distribution Network: DHS relies on a network of insulated pipes to distribute hot water, steam, or other heat carriers from the central plants to individual buildings or facilities within the district. This distribution network ensures the efficient and reliable delivery of thermal energy to consumers.
- Multiple Heat Sources: DHS can utilize a variety of heat sources to generate thermal energy, including CHP plants, geothermal energy, biomass boilers, waste-to-energy plants, solar thermal collectors, and waste heat recovery systems. This diversity of heat sources enhances the resilience and sustainability of the system.
- Energy Efficiency: By centralizing heat production and distribution, DHS achieve higher levels of energy efficiency compared to decentralized heating systems. This efficiency is further enhanced through the use of CHP technology, which simultaneously produces electricity and heat from a single fuel source.
- Integration with Renewable Energy: DHS can easily integrate renewable energy sources such as solar thermal, geothermal, and biomass into the system. This allows for the reduction of greenhouse gas emissions and promotes the transition to low-carbon energy sources.
- Reliability and Resilience: DHS offers greater reliability and resilience compared to decentralized heating systems. With redundant infrastructure and backup systems in place, DHS can continue to

provide heat even during equipment failures or extreme weather events, ensuring uninterrupted service for consumers.

- Flexibility and Scalability: DHS is highly flexible and scalable, allowing for easy expansion or modification to accommodate changes in heating demand or the integration of new heat sources. This flexibility makes DHS suitable for a wide range of applications, from small-scale residential developments to large industrial complexes.
- Environmental Sustainability: DHS promotes environmental sustainability by reducing greenhouse gas emissions, improving air quality, and minimizing the consumption of finite resources. By utilizing cleaner and more efficient energy sources, DHS contributes to the mitigation of climate change and the preservation of natural ecosystems.

1.1.3 Definition of District Cooling Systems (DCS)

Similar to DHS described in Section 1.1.1, a district cooling system (DCS), a centralized cooling network, is designed to deliver chilled water or other cooling mediums from a central plant to multiple buildings within a designated area or district. This system operates by generating cooling at a central chiller plant, where large-scale chillers and cooling towers produce chilled water, which is then distributed through an insulated piping network to connected buildings through energy transfer stations (ETS). These buildings use the chilled water for air conditioning, refrigeration, and other cooling needs.

DCS provides a more efficient, reliable, and environmentally friendly alternative to traditional individual air-conditioning units. By consolidating cooling production, DCS can achieve higher energy efficiency, reduce operational costs, and lower greenhouse gas emissions. Additionally, DCS can integrate various energy sources, including renewable energy, waste heat recovery, and advanced technologies, further enhancing their sustainability and performance.

1.1.3.1 Key Features of District Cooling Systems (DCS)

DCS is distinguished by several key features that collectively make them efficient, sustainable, and reliable solutions for urban cooling needs. At the core of a DCS is the centralized chilling plant, which utilizes large-scale chillers, such as electric-driven, absorption, and centrifugal chillers, to produce chilled water. This chilled water is then distributed through an insulated network of pipes to connected buildings, minimizing thermal losses and ensuring efficient delivery of cooling energy. Each building in the network is equipped with an energy transfer station that uses heat exchangers to transfer the cooling energy from the chilled water to the building's internal system, providing consistent and efficient cooling. The system operates on a closed loop, with warmed water returning to the central plant to be re-cooled and re-circulated. DCS can integrate multiple cooling sources, including renewable energy options like solar thermal and geothermal, enhancing its energy efficiency and environmental sustainability. This centralized approach results in significant cost savings due to economies of scale and reduced maintenance requirements, while also lowering greenhouse gas emissions and improving air quality. Furthermore, DCS offers enhanced reliability and resilience, with built-in redundancies and backup systems ensuring uninterrupted operation. The system is also highly flexible and scalable, capable of expanding to meet growing demand and incorporating new technologies. By centralizing cooling infrastructure, DCS frees up valuable space within buildings, reduces noise pollution, and supports better urban planning and aesthetics. Additionally, the system effectively manages peak cooling loads, optimizing resource use and improving overall efficiency. These features make DCS an ideal solution for urban areas, commercial complexes, and industrial facilities seeking sustainable and reliable cooling solutions.

1.2 Historical Development and Evolution of District Heating and Cooling Systems (DCHS)

The evolution of DCHS has been a journey marked by innovation, technological advancements, and a growing emphasis on sustainability. Understanding the historical context of these systems provides valuable insights into their current applications and future potential.

1.2.1 Early Beginnings and Development

The concept of district heating can be traced back to ancient civilizations. The Romans, known for their engineering prowess, developed early forms of district heating through hypocaust systems, which provided central heating to public baths and buildings using hot air and smoke channeled through walls and floors. This rudimentary form of district heating laid the foundation for future developments. In the 14th century, the city of Chaudes-Aigues in France pioneered the use of geothermal energy for district heating, utilizing natural hot springs to heat buildings. This marked one of the earliest known applications of geothermal energy for communal heating.

1.2.2 Industrial Revolution and Modernization

The Industrial Revolution in the 19th century brought significant advancements to district heating. In 1877, Birdsill Holly established the first modern DHS in Lockport, New York, using steam generated from a central plant to heat multiple buildings via a network of pipes. This innovation set the stage for widespread adoption in urban centers, particularly, in Europe and North America. By the early 20th century, major cities like New York, Paris, and Copenhagen developed extensive district heating networks. These systems primarily relied on coal-fired boilers and later transitioned to oil and natural gas, reflecting the evolving energy landscape of the time.

1.2.3 Post-World War II Expansion

The post-World War II era saw substantial growth in DHS, driven by reconstruction efforts and rapid urbanization. Europe, especially, the Nordic countries—Denmark, Sweden, and Finland—emerged as leaders in advanced DHS. These countries integrated CHP plants and utilized waste heat from industrial processes, significantly improving energy efficiency and reducing emissions.

1.2.4 Emergence of District Cooling Systems

While DHS had established a strong foothold, DCS began to gain prominence in the latter half of the 20th century. The increasing demand for air conditioning in urban environments necessitated centralized cooling solutions. In the 1960s and 1970s, cities in the Middle East, such as Dubai and Abu Dhabi, adopted DCS to combat extreme heat and support urban growth. These systems employed large-scale chillers and cooling towers to produce chilled water, which was then distributed to buildings for air conditioning.

1.2.5 Technological Advancements and Integration

Advancements in DCHS and DCS technology burst through at the end of the 20th and beginning of the 21st century. Improvements in energy efficiency and digitally enabled control technologies were augmented by power generated from renewable energy. By the early 2000s, DHS had started to incorporate renewable energy produced by heat pumps, biomass, geothermal, and solar thermal sources. Smart grid technologies and advanced control systems were also starting to be included in DCHS, which led to greater optimization and management of demand.

Likewise, DCS promoted sustainability, combining renewable energy and cutting-edge cooling technologies with the use of seawater and lake water for cooling, as well as free cooling, which significantly reduced energy consumption and, therefore, had less environmental impact.

1.3 Different Generations of District Cooling Systems (DCS)

Despite various evolutions of DCS, there was no categorization of generations of DCS prior to 2014. It is paramount to visualize the trends of various DCS that have evolved. Thus, industry and researchers have created a common terminology and framework for DCS, resulting in a useful benchmark for both historical and future perspectives of DCS systems. Therefore, various generations of DCS are briefly discussed in Sections 1.3.1–1.3.5.

1.3.1 First Generation of District Cooling Systems (1GDC)

In the late 19th century, there was a growing demand for cold energy storage in the food supply chain in the United States. This led to the development of communal pipeline refrigeration systems, which marked the 1GDC that utilizes centralized condensers to connect many evaporators, whether they are centralized or decentralized [1].

In 1GDC, buildings are interconnected within a network. This allows for precise measurement of cooling consumption across the entire network and individual buildings. One key strategy involves optimizing the use of chillers. Surveys indicate that installed cooling capacity often exceeds actual demand, with a significant portion being inactive spare capacity. By analyzing a year of operation, it is possible to determine the necessary capacity and identify which plants can be decommissioned. This approach not only improves efficiency but also reduces investment costs in cooling capacity. Large central chillers, commonly used in the first generation, can be further enhanced to achieve optimal performance. Despite improving the efficiency of 1GDC, there were several weaknesses as follows:

- **Centralization**: The reliance on large central chillers can lead to a single point of failure. If the central plant experiences an issue, it affects the entire network.
- **Inflexibility**: The first-generation systems may lack flexibility in adjusting to varying cooling demands. They often operate at a fixed capacity, which can result in inefficiencies during low-demand periods.
- **High Initial Costs**: Establishing a district cooling network involves significant upfront investment in infrastructure, including piping, chillers, and distribution systems.
- **Maintenance Challenges**: Maintaining large central chillers and the interconnected network can be complex and costly. Regular maintenance is crucial to prevent breakdowns.

- **Energy Loss**: Despite being more efficient than individual building-level chillers, the transmission of chilled water through underground pipes can still result in energy losses.

Efforts to address these weaknesses have led to advancements in the second generation of district cooling systems (2GDC).

1.3.2 Second Generation of District Cooling Systems (2GDC)

In 2GDC, thermal energy storage (TES) for cooling and chilled water storage facilities play a crucial role. These facilities offer several advantages, including reduced power consumption and better management of peak demand. Here's how it works: during off-peak hours when electricity rates are lower, chilled water is stored. Then, during peak demand periods, this stored chilled water is used, effectively shifting the load. Chilled water storage aligns well with existing chiller product lines, requiring minimal modifications. However, a challenge lies in maintaining supply and return separation to maximize capacity and maintain cool supply temperatures. Overall, the second generation leverages chilled water storage to enhance efficiency and address peak demand challenges.

Thermal storage systems deployed in 2GDC, particularly chilled water storage, have their own set of drawbacks. The demand for a huge physical site space posed significant challenges in deploying 2GDC in densely populated urban areas. Furthermore, storage infrastructures, such as tanks, pumps, and associated equipment, require a huge upfront cost. Moreover, well-maintained thermal storage systems and associated equipment are essential to ensure optimal performance. Since thermal storage systems are equipped in 2GDC, ensuring uniform temperature distribution within storage tanks is challenging, affecting overall system efficiency. To mitigate several drawbacks, industry and researchers worked on exploring the third generation of district cooling systems (3GDC).

1.3.3 Third Generation (3GDC Late 20th Century—Present)

3GDC focuses on enhancing efficiency, particularly when there's a demand for heating in buildings, hot tap water, or industrial processes. These systems integrate a variety of cooling sources, including absorption chillers, heat pumps, and mechanical chillers, enhancing their adaptability and efficiency. They are often coupled with DHS and utilize short-term cold storage to optimize energy usage. By leveraging natural cooling sources such as rivers, lakes, and seawater, alongside waste cold streams from gas expansion, third-generation systems provide a sustainable and cost-effective cooling solution. Table 1.1 summarizes some of the notable 3GDC, technology deployed for cold energy, and their cold supply capacity.

TABLE 1.1
Notable Third-Generation District Cooling Systems

Location	Starting Year	Cold Supply Technology	Cold Capacity	Application
Stockholm, Sweden	1995	Compression and absorption chillers, seawater cooling, heat pumps	335 GWh/year	Commercial buildings
Gothenburg, Sweden	1995	Compression and absorption chillers; river cooling	70 MW	Commercial buildings
Helsinki, Finland	1998	Heat pump, absorption chillers, and sea water cooling	200 MW	Commercial buildings
Toronto, Canada	2004	Natural cooling, compression, and absorption chillers	260 MW	Commercial buildings
Dubai, UAE	2003	Electric chillers, renewable energy integration	1,530,000 RT	Residential, commercial, and mixed-use buildings
Marina Bay, Singapore	2006	Chilled water technology	320 MW	Commercial buildings
Kai Tak Development, Hong Kong	2013	Centralized cooling plant	210 MW	Residential and commercial buildings

1.3.4 Fourth Generation (4GDC Emerging Trends)

Previous DC generations focus on supplying the needs for cooling in districts, enabling energy-efficiency improvement. The rapid advancement in smart technologies pushes DCS beyond satisfying thermal loads in buildings. The evolution of the fourth-generation district cooling systems (4GDC) enables seamless integration with electricity, gas, and cooling systems. 4GDC utilizes multiple cold sources, including renewable energies like solar, wind, and geothermal, as well as natural cooling from rivers, lakes, and seawater. They also incorporate waste cold streams and renewable electricity. Operating efficiently at higher temperatures to reduce energy consumption, they feature advanced thermal storage for both hot and cold storage needs. The inclusion of advanced digital controls allows for optimized performance through real-time monitoring, predictive maintenance, and load management. This results in improved energy system efficiency, reduced greenhouse gas emissions, and enhanced urban resilience. Overall, 4GDC are designed to provide sustainable and cost-effective cooling solutions for residential, commercial, and mixed-use developments.

1.3.5 Fifth Generation (Future Possibilities)

Despite focusing on energy efficiency in 4GDC, fifth-generation district cooling systems (5GDC) are an emerging concept that pushes the boundaries of sustainability, efficiency, and integration beyond the advancements made in the fourth generation. These systems are designed to be even more adaptive, leveraging cutting-edge technologies and innovative approaches to meet the evolving needs of modern urban environments. 5GDC epitomize the future of urban cooling with a focus on ultra-low temperature distribution to enhance efficiency and reduce thermal losses. These systems seamlessly integrate with renewable energy sources, such as solar, wind, geothermal, and biomass, and utilize advanced energy storage solutions to balance supply and demand effectively. Their decentralized and modular design ensures flexibility, scalability, and resilience by employing multiple smaller cooling units distributed throughout the network. Advanced digitalization through artificial intelligence, machine learning, and Internet of Things (IoT) enables predictive maintenance, real-time optimization, and enhanced energy management, while smart sensors and data analytics provide deep insights into system performance. Emphasizing sustainability, these systems maximize heat recovery from various sources for reuse in district heating or hot water supply, aiming for carbon neutrality and contributing to green city initiatives. Enhanced user interaction through user-friendly interfaces and mobile applications offers greater control and customization of cooling needs. Additionally, a collaborative and community-based approach involving utilities, local governments, businesses, and residents promotes shared infrastructure and optimized resource use, ensuring economic viability and a sustainable future for urban cooling.

The potential benefits of 5GDC are substantial, starting with significant energy-efficiency gains through ultra-low temperature distribution and advanced energy management. This results in considerable energy savings and reduced operational costs. Their decentralized and modular design enhances the system's flexibility and resilience, allowing for quick adaptation to changing demands and rapid recovery from disruptions. By extensively utilizing renewable energy and maximizing heat recovery, these systems contribute to carbon neutrality and minimize environmental impact, aligning with global sustainability goals. Enhanced user experience is another key benefit, as advanced digitalization and user-friendly interfaces provide end-users with greater control and convenience. Furthermore, the collaborative and community-based approach fosters partnerships among utilities, local governments, businesses, and residents, optimizing resource use and lowering costs through shared infrastructure. Collectively, these advantages position 5GDC as a forward-thinking solution that meets the evolving needs of modern urban environments while promoting economic, environmental, and social sustainability.

1.4 Concept of District Energy Sharing

As urbanization accelerates and the demand for efficient and sustainable cooling solutions intensifies, the concept of energy sharing in DCS emerges as a promising approach to enhance overall system efficiency and operational flexibility. Traditional DCS often face challenges such as energy losses, high initial costs, and limited adaptability to varying demands. Energy sharing addresses these issues by creating an interconnected network that optimizes the use of available thermal energy resources. By integrating diverse energy sources, including renewables and waste heat, and leveraging advanced digital technologies, energy sharing maximizes energy efficiency, reduces operational costs, and enhances the resilience of cooling systems. This innovative approach not only meets the cooling needs of modern urban environments but also aligns with global sustainability goals.

Energy sharing in DCS represents a transformative approach to urban energy management, focusing on the efficient distribution and utilization of thermal energy across multiple buildings and facilities. The core idea is to create an interconnected network where heating and cooling demands are balanced through the sharing of energy resources. This involves integrating various energy sources, including renewable energy, waste heat, and natural cooling, to optimize the overall energy efficiency of the system. By leveraging advanced technologies such as heat pumps, absorption chillers, and thermal storage, energy-sharing systems can dynamically adjust to the fluctuating demands of different buildings, ensuring that excess heat or cooling from one location can be redirected to another where it is needed.

One of the key features of energy sharing in DCS is the extensive use of renewable energy sources and waste energy recovery. Renewable energy sources like solar thermal, geothermal, and biomass are integrated into the cooling network to provide a sustainable supply of thermal energy. Additionally, waste heat from industrial processes, data centers, and other sources is captured and repurposed within the network. This not only enhances the overall efficiency of the system but also reduces the dependency on fossil fuels, thereby lowering greenhouse gas emissions. By converting what would otherwise be wasted energy into useful thermal energy, DCS can achieve significant environmental benefits.

The incorporation of advanced digital technologies and smart grids is essential for the effective operation of energy-sharing systems. Smart grids enable real-time monitoring and control of energy flows within the network, allowing for precise management of thermal energy distribution. IoT devices and sensors provide valuable data on system performance and user demand, which can be analyzed using machine learning algorithms to optimize energy usage. Predictive maintenance and load management capabilities further enhance the reliability and efficiency of the system. By using digitalization, energy-sharing systems can anticipate demand patterns, reduce energy wastage, and ensure that cooling is provided where and when it is most needed.

1.4.1 DCHS Market

The market for DCHS is experiencing robust growth globally, driven by the increasing demand for energy-efficient and sustainable urban infrastructure. Rapid urbanization, coupled with the need to mitigate the environmental impact of traditional heating and cooling methods, is propelling the adoption of DCHS in both developed and emerging economies. Governments and municipalities are increasingly investing in DCHS as part of smart city initiatives, aiming to enhance energy efficiency, reduce greenhouse gas emissions, and improve urban resilience.

1.4.2 Current Market Size and Growth

1.4.2.1 Market Drivers and Regional Insights

Several factors contribute to the expanding DCHS market. First, technological advancements, such as the integration of renewable energy sources and smart grid technologies, are enhancing the efficiency and appeal of DCHS. The ability to utilize waste heat from industrial processes and incorporate renewable energy sources like solar, geothermal, and biomass significantly reduces the carbon footprint of heating and cooling operations. Furthermore, regulatory policies promoting energy efficiency and sustainability, along with financial incentives for green building projects, are encouraging the deployment of DCHS.

The Middle East and Africa region [9], particularly countries like the UAE, Saudi Arabia, and Qatar, lead the global district cooling market. This region experiences extreme temperatures, creating a high demand for efficient and reliable cooling solutions. Significant investments in infrastructure and urban development projects further drive market growth. In the UAE, major players such as Empower in Dubai [10], with a plant capacity of 1.53 million RT, and Qatar Cool [11] in Doha, with a capacity of 157,000 RT, are spearheading the adoption of DCS, which are becoming integral to urban planning and sustainability efforts. Saudi Arabia's Saudi Tabreed, with a substantial plant capacity, also plays a crucial role, contributing to the region's dominance in the district cooling market.

The Asia Pacific region [9] is witnessing rapid growth in the district cooling market, driven by urbanization, industrialization, and supportive government policies promoting sustainable development. Countries like China, India, and Singapore are at the forefront, implementing large-scale district cooling projects to enhance energy efficiency and reduce environmental impact. Singapore's Marina Bay DCS, with a capacity of 320 MW, and Hong Kong's Kai Tak Development, with a capacity of 284 MW [12], are prime examples of successful implementations. Additionally, Malaysia's Cyberjaya and Putrajaya projects, with capacities of 60 and 135 MW, respectively, highlight the region's commitment to integrating district cooling into urban infrastructure to meet the growing demand for efficient cooling solutions.

In North America [9], the district cooling market is driven by the need for energy-efficient and sustainable cooling solutions in urban areas. The region sees high adoption rates in commercial and institutional buildings, supported by technological advancements and stringent energy regulations. Notable projects include Con Edison's district cooling system in New York, with a capacity of 30,000 RT, Cornell University's innovative cooling solutions with a capacity of 70 MW, and Toronto's Deep Lake Water Cooling project in Canada, with a capacity of 264 MW. These projects demonstrate the region's focus on leveraging advanced technologies to provide reliable and sustainable cooling services, reducing energy consumption and environmental impact.

Europe [9] is experiencing significant growth in the district cooling market due to stringent environmental regulations and a strong emphasis on sustainability. The integration of renewable energy sources and advanced district energy systems is a key driver in this region. Leading examples include Climespace in Paris, France, with a capacity of 472 MW, Stockholm Exergi in Sweden, with a capacity of 335 GWh/year, and Helsinki's district cooling network in Finland, with a capacity of 200 MW. These systems exemplify Europe's commitment to reducing carbon emissions and enhancing energy efficiency. The region's focus on green urban planning and adherence to strict environmental standards are pivotal in advancing the adoption of district cooling solutions.

Overall, the DCHS market is poised for significant growth, supported by a global shift toward sustainable urban development and the continuous evolution of technology. As more cities adopt DCHS, the benefits of reduced operational costs, improved energy efficiency, and lower environmental impact will become increasingly apparent, driving further investment and innovation in this vital sector.

1.4.2.2 Market Challenges

Despite rapid growth in DCS system deployments in various regions, there are several challenges encountered in the DCS market, from high initial costs and regulatory complexities. Therefore, challenges faced in the DCS market are highly dependent on the regions.

In the Middle East and Africa, the district cooling market faces several challenges despite its growth potential. One significant issue is the high initial capital investment required for infrastructure development. Establishing centralized cooling plants and extensive distribution networks demands substantial financial resources, which can be a barrier for many projects. Additionally, the region's extreme climate conditions necessitate robust and reliable cooling systems, which further increases costs. Regulatory and policy frameworks in some countries may not be fully supportive or harmonized, leading to bureaucratic delays and operational inefficiencies. Furthermore, the region's reliance on conventional energy sources means that integrating renewable energy into DCS is often limited, posing sustainability challenges.

The Asia Pacific region, while experiencing rapid growth in the district cooling market, faces challenges related to rapid urbanization and industrialization. The rapid pace of development can strain existing infrastructure, making it difficult to keep up with the growing demand for cooling. High upfront costs for establishing district cooling networks can be prohibitive, especially in developing countries. Additionally, the region's diverse regulatory environments and varying levels of policy support for sustainable energy solutions can create uncertainties and hinder market expansion. Integrating DCS with renewable energy sources remains a challenge due to technological and financial constraints. Furthermore, public awareness and acceptance of district cooling solutions are still developing, which can affect market penetration.

In North America, the district cooling market encounters challenges related to regulatory complexities and market fragmentation. Different states and municipalities have varying regulations and standards, making it difficult to implement uniform district cooling solutions. The high initial costs of installing district cooling infrastructure and retrofitting existing buildings can also be a significant barrier. Moreover, there is often a lack of incentives and subsidies for district cooling projects compared to other renewable energy initiatives. The integration of advanced technologies, such as smart grids and IoT, while beneficial, requires significant investment and can pose cybersecurity risks. Additionally, the market is highly competitive, with traditional heating ventilation and air conditioning (HVAC) systems still being widely used, which can limit the adoption of district cooling solutions.

In Europe, the district cooling market faces challenges primarily related to stringent environmental regulations and the need for sustainable solutions. While these regulations drive the adoption of district cooling, they also require significant compliance efforts and investments in advanced technologies. High installation and operational costs are major barriers, particularly in historic cities where integrating modern infrastructure can be complex and costly. The transition to renewable energy sources and the integration of district cooling with existing DHS require substantial technological advancements and financial support. Additionally, public awareness and acceptance of DCS vary across countries, affecting market growth. Fragmented regulatory frameworks within the European Union can also lead to inconsistencies and delays in project implementation.

1.4.3 Advancements in District Cooling Systems

DCS have seen significant technological advancements over the years, enhancing their efficiency, sustainability, and integration with modern urban infrastructure. These advancements are pivotal in addressing the challenges of energy consumption, environmental impact, and operational efficiency.

1.4.3.1 Advanced Chillers

Modern DCS incorporate advanced chiller technologies, including electric, absorption, and centrifugal chillers. These chillers are designed to operate with higher efficiency, lower energy consumption, and reduced environmental impact. Variable speed drives (VSDs) and advanced control systems optimize the operation of chillers, adjusting their output based on real-time demand and thereby improving energy efficiency.

1.4.3.2 Thermal Energy Storage

TES systems are a significant advancement in district cooling. TES allows for the storage of chilled water or ice during off-peak hours, which can then be used during peak demand periods. This not only reduces peak electricity demand but also enhances the overall efficiency of the cooling system. TES systems are particularly beneficial in regions with high variations in energy demand and supply, providing a reliable cooling solution.

1.4.3.3 Renewable Energy Integration

The integration of renewable energy sources, such as solar thermal, geothermal, and biomass, into DCS represents a major technological leap. Solar cooling systems use solar energy to drive absorption chillers, reducing reliance on conventional energy sources. Geothermal systems exploit the earth's constant temperatures to provide efficient cooling, while biomass systems use organic materials to generate energy for cooling. These renewable integrations help in reducing carbon emissions and promoting sustainable energy use.

1.4.3.4 Smart Grids and IoT

The adoption of smart grid technologies and the IoT has revolutionized DCS. Smart grids enable real-time monitoring and control of energy distribution, optimizing the operation of the entire network. IoT devices and sensors collect data on temperature, humidity, and energy consumption, which is then analyzed to improve system performance. Machine learning algorithms can predict cooling demand patterns, allowing for proactive adjustments and efficient energy management.

1.4.3.5 Advanced Control Systems

Sophisticated control systems play a crucial role in modern district cooling. These systems use predictive analytics and artificial intelligence to manage and optimize cooling operations. For example, demand response strategies adjust cooling loads based on real-time energy prices and grid conditions. Building management systems (BMS) integrate with district cooling networks to provide seamless control and monitoring, enhancing the overall efficiency and reliability of the cooling service.

1.4.3.6 High-Efficiency Pumps and Distribution Networks

Technological advancements in pump design and distribution networks have significantly improved the efficiency of DCS. High-efficiency pumps reduce energy consumption and operational costs. Advanced pipe materials and insulation techniques minimize thermal losses during the distribution of chilled water. Additionally, the use of hydraulic modeling and simulation tools helps in designing optimized distribution networks that ensure efficient energy transfer.

1.4.3.7 Waste Heat Recovery

The incorporation of waste heat recovery systems in district cooling networks is another technological advancement. These systems capture waste heat from industrial processes, power plants, and data centers, converting it into useful thermal energy for cooling. This not only enhances the energy efficiency of the cooling system but also contributes to overall energy savings and environmental sustainability.

1.5 The Importance of District Cooling and Heating Systems in Sustainability

DCHS play a crucial role in promoting sustainability in urban environments. These systems are designed to provide efficient, centralized thermal energy distribution for both cooling and heating, significantly reducing the environmental impact compared to traditional, decentralized methods.

1.5.1 Energy Efficiency

One of the primary advantages of DCHS is their superior energy efficiency. By centralizing the production of thermal energy, these systems can achieve higher efficiencies than individual building systems. Centralized plants can optimize their operations, using advanced technologies like VSDs, efficient chillers, and CHP systems to minimize energy consumption. This optimization results in lower overall energy use and reduced greenhouse gas emissions, contributing to a more sustainable urban energy landscape.

1.5.2 Integration with Renewable Energy

DCHS are well-suited for integrating renewable energy sources, which further enhances their sustainability credentials. Solar thermal energy, geothermal energy, and biomass can be effectively utilized within these systems. For instance, solar thermal collectors can provide hot water for district heating, while geothermal systems can offer both heating and cooling solutions. The use of renewable energy reduces reliance on fossil fuels, decreases carbon emissions, and helps mitigate climate change.

1.5.3 Reduction of Carbon Footprint

DCHS significantly reduce the carbon footprint of urban areas. By utilizing waste heat from industrial processes, power plants, or data centers, these systems convert what would be lost energy into useful thermal energy. This waste heat recovery not only enhances energy efficiency but also lowers the carbon emissions associated with traditional heating and cooling methods. Additionally, the centralized nature of DCHS allows for better monitoring and control, ensuring that emissions are minimized.

1.5.4 Economic Benefits

The economic benefits of DCHS contribute to their sustainability. Centralized systems can take advantage of economies of scale, reducing the cost of energy production and distribution. Lower operational costs translate to cost savings for consumers, making these systems economically viable. Moreover, the implementation of DCHS can attract investment in green infrastructure, creating jobs and boosting local economies. These economic incentives support the long-term sustainability of urban areas.

1.5.5 Enhanced Urban Resilience

DCHS enhance the resilience of urban energy systems. By providing a reliable and stable supply of heating and cooling, these systems help cities withstand extreme weather conditions and energy supply disruptions. The ability to integrate diverse energy sources, including renewables and waste heat, ensures a consistent and resilient energy supply. This resilience is crucial for maintaining urban sustainability in the face of climate change and other challenges.

1.5.6 Improved Air Quality

The use of DCHS can lead to improved air quality in urban areas. Traditional heating methods, such as individual boilers, often rely on fossil fuels, emitting pollutants like nitrogen oxides and particulate matter. By centralizing heating and using cleaner energy sources, DCHS reduce these harmful emissions, resulting in cleaner air and better public health outcomes. Improved air quality is a key aspect of urban sustainability, enhancing the quality of life for city residents.

1.5.7 Urban Planning and Space Efficiency

DCHS contribute to more efficient urban planning and space utilization. By eliminating the need for individual heating and cooling units in each building, these systems free up valuable space that can be used for other purposes. This efficiency is particularly important in densely populated urban areas where space is at a premium. Efficient use of space supports sustainable urban development by allowing for more green spaces, public amenities, and efficient infrastructure planning.

1.6 Conclusion

In summary, DCHS are essential components of modern urban infrastructure, offering sustainable, energy-efficient solutions for meeting the growing demands of cooling and heating in cities worldwide. These systems have evolved through several generations, from early pipeline refrigeration to the latest innovations integrating renewable energy sources and advanced digital technologies. As highlighted in this chapter, DCHS not only improve energy efficiency and operational costs through centralized production and distribution of thermal energy but also contribute significantly to carbon reduction efforts and environmental sustainability. The integration of renewable energy, waste heat recovery, and smart grid technologies has further enhanced their performance, making them crucial for reducing the carbon footprint of urban environments. Despite challenges such as high initial costs and regional regulatory complexities, the advancements in technology and the growing adoption of DCHS globally signal a promising future. As cities continue to prioritize sustainability, energy efficiency, and resilience, DCHS will play a pivotal role in shaping the energy landscape of urban environments, driving the transition toward a low-carbon, sustainable future.

References

[1] Østergaard, P. A., Werner, S., Dyrelund, A., Lund, H., Arabkoohsar, A., Sorknæs, P., Gudmundsson, O., Thorsen, J. E., & Mathiesen, B. V. (2022). The four generations of district cooling – A categorization of the development in district cooling from origin to future prospect. *Energy*, 253, 124098. https://doi.org/10.1016/j.energy.2022.124098

[2] L. Minh Dang, Le Quan Nguyen, Junyoung Nam, Nguyen, T. N., Sujin Lee, Hyoung-Kyu Song, Hyeonjoon Moon (2024). Fifth generation district heating and cooling: A comprehensive survey. *Energy Reports*, 11, 1723–1741. https://doi.org/10.1016/j.egyr.2024.01.037

[3] Yong-Hoon, Im, & Jie, Liu (2018). Feasibility study on the low-temperature district heating and cooling system with bi-lateral heat trades model, *Energy*, 153, 988–999. https://doi.org/10.1016/j.energy.2018.04.094

[4] Schmidt, D., Kallert, A., Blesl, M., Svendsen, S., Hongwei Li, Nord, N., & Sipilä, K. (2017). Low temperature district heating for future energy systems. *Energy Procedia*, 116, 26–38. https://doi.org/10.1016/j.egypro.2017.05.052

[5] Ommen, T., Thorsen, J. E., Markussen, W. B., & Elmegaard, B. (2017). Performance of ultra low temperature district heating systems with utility plant and booster heat pumps. *Energy*, 137, 544–555. https://doi.org/10.1016/j.energy.2017.05.165

[6] Bünning, Felix, Wetter, Michael, Fuchs, Marcus, & Müller, Dirk (2018). Bidirectional low temperature district energy systems with agent-based control: Performance comparison and operation optimization. *Applied Energy*, 209, 502–515. https://doi.org/10.1016/j.apenergy.2017.10.072

[7] Pellegrini, M., & Bianchini, A. (2018). The innovative concept of cold district heating networks: A literature review. *Energies*, 11, 236. https://doi.org/10.3390/en11010236

[8] Ahmed Abdalla, Saber Mohamed, Scott Bucking, James S. Cotton (2021). Modeling of thermal energy sharing in integrated energy communities with micro-thermal networks. *Energy and Buildings*, 248, 111170. https://doi.org/10.1016/j.enbuild.2021.111170

[9] Sorour Alotaibi, Mohammad Alhuyi Nazari (2023). District cooling in the Middle East & North Africa; history, current status, and future opportunities. *Journal of Building Engineering*, 77, 107522. https://doi.org/10.1016/j.jobe.2023.107522

[10] https://www.emicool.com/en/projects (last accessed on August 01, 2024)

[11] https://www.qatarcool.com/qa/about-us/our-districts/ (last accessed on August 01, 2024)

[12] https://www.emsd.gov.hk/en/energy_efficiency/district_cooling_system/introduction/index.html (last accessed on August 01, 2024)

2

Central Air-Conditioning Chiller System

2.1 Working Principle of Central Air-Conditioning Chiller Plant

Water and air-cooled central chiller systems, water-cooled package units, stand-alone package units, and variable refrigerant volume (VRV) systems are commonly used to maintain the comfort condition of buildings and industrial plants. Out of these systems, the most widely used system in large buildings, industrial plants, and district cooling systems is the water-cooled central chiller system [1]. The major components of the water-cooled central chiller system are vapor compression type chiller, chilled water pump, condenser water pump, cooling tower, air handling unit (AHU), blower, flow-modulating valve, and energy management system as shown in Figure 2.1. The energy management system, which is not shown in Figure 2.1, controls the operation of each component based on the dynamic cooling loads of the air-conditioned spaces.

A vapor compression type chiller contains four main components, namely, the compression, condenser, expansion device, and evaporator as illustrated in Figure 2.2 [2]. Chillers are cyclic devices. The working fluids used in the chiller cycles are called refrigerants. The desirable characteristics of a refrigerant include high enthalpy of vaporization, phase change temperatures from liquid to vapor (in the evaporator) and vapor to liquid (in the condenser) and the corresponding pressures, ozone depletion potential, global warming potential, toxicity, corrosiveness, flammability, chemical stability, and cost. Chlorofluorocarbons (CFCs), hydrofluorocarbons (HFCs), hydrochlorofluorocarbons (HCFCs), ammonia, and hydrocarbons are commonly used refrigerants. The ozone crisis has triggered a critical evaluation and selection of refrigerants. CFCs allow more ultraviolet radiation into the Earth's atmosphere by destroying the protective ozone layer. HCFCs have about 5% of the ozone depletion potential of CFCs. Unlike CFCs and HCFCs, HFCs do not contribute to ozone depletion. R-134a, R-410A, R-407C (all are HFCs), and R-502 (a blend of R115 and R-22) are among the dominant refrigerants used in commercial chiller systems.

DOI: 10.1201/9781032670164-2

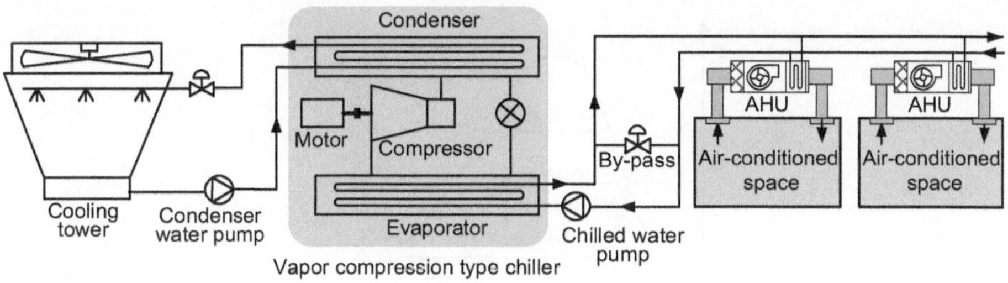

FIGURE 2.1
Water-cooled central chiller system.

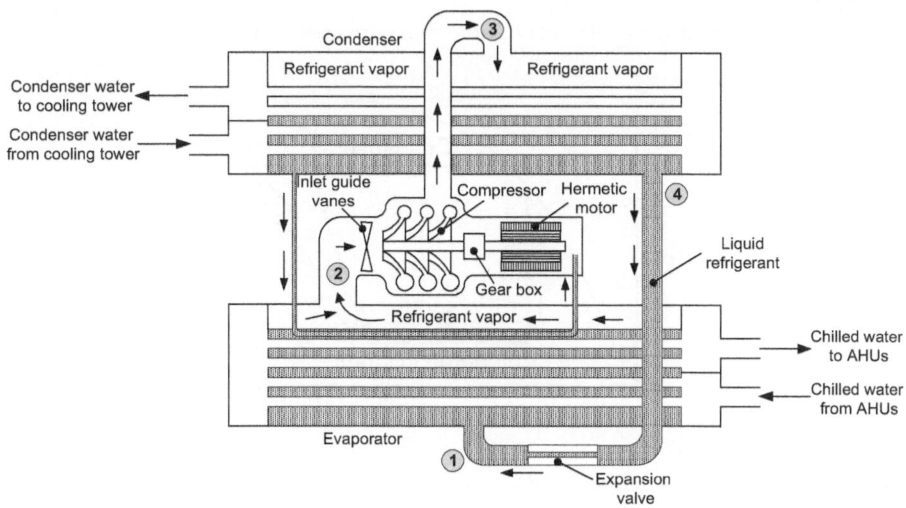

FIGURE 2.2
Vapor compression type water-cooled chiller.

The temperature–entropy diagram and the pressure–enthalpy diagram of the vapor compression refrigeration cycle are shown in Figures 2.3 and 2.4, respectively. The phase change temperature of a fluid decreases with the decrease of its pressure. The low-pressure liquid-vapor mixture of refrigerant enters the evaporator at State 1. The refrigerant absorbs the heat of vaporization from the tube bank of the evaporator and leaves the evaporator as a saturated or slightly superheated vapor at State 2. The chilled water flowing through the tube bank of the evaporator rejects heat to the refrigerant and leaves the tube bank as low-temperature chilled water, which is supplied to the cooling coils of the AHUs, as illustrated in Figure 2.1. The saturated or slightly superheated refrigerant vapor enters the compressor at State 2. The compressor increases the pressure of the refrigerant to the condenser pressure by absorbing the electric energy. The temperature of the refrigerant

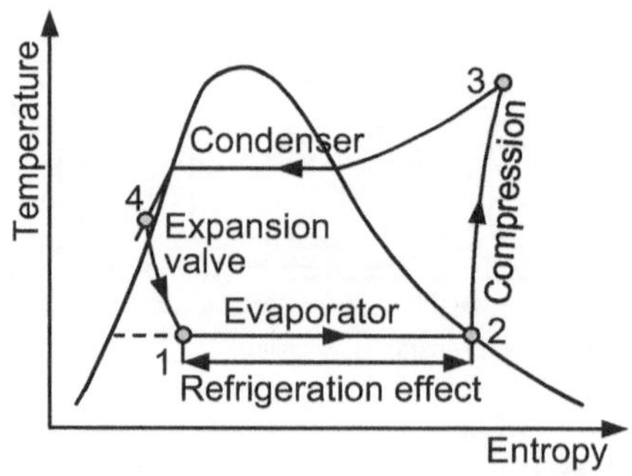

FIGURE 2.3
Temperature–entropy diagram of vapor compression refrigeration cycle.

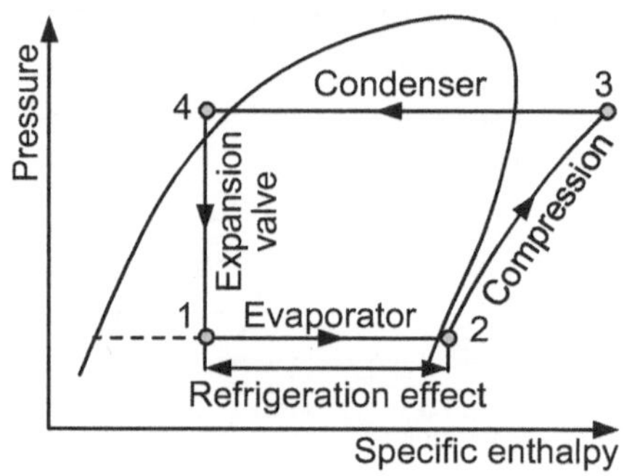

FIGURE 2.4
Pressure-specific enthalpy diagram of vapor compression refrigeration cycle.

increases during the compression process to well above the temperature of the condenser water. The refrigerant then enters the condenser as superheated vapor at State 3 and leaves as a saturated or slightly subcooled liquid at State 4 because of heat rejection to the condenser water flowing through the tube bank of the condenser. The warm condenser water is circulated through the cooling tower to reject the absorbed heat to the ambient air. The saturated or slightly subcooled liquid refrigerant at State 4 is throttled to the evaporator pressure by passing through an expansion device, such as an expansion valve or orifice plate, as shown in Figure 2.2. Due to the reduction of pressure, the boiling point of the liquid refrigerant drops. A small fraction of the refrigerant evaporates by absorbing the heat of vaporization from the liquid refrigerant itself. Hence, the temperature of the refrigerant drops and leaves the expansion valve as a low-temperature liquid-vapor mixture at State 1 and re-enters the evaporator, completing the cycle. To have heat transfer at a reasonable rate, a temperature difference of about 5°C is maintained between the refrigerant and the inlet water with which it is exchanging heat.

The difference between the temperature of chilled water leaving the evaporator and the temperature of the refrigerant entering the evaporator is known as chilled water approach temperature. Similarly, the difference between the temperature of the refrigerant entering the condenser and the temperature of condenser water leaving the condenser is known as the condenser water approach temperature. An approach temperature of about 1°C is achieved if the tubes of the evaporator and condenser are kept clean. A higher approach temperature indicates the accumulation of duct or scale formation on the tube surfaces, which hinders the heat transfer rate between the refrigerant and water, resulting in a higher differential pressure across the compressor and drops in energy performance of the chiller.

The cooling effect produced by a chiller can be calculated in refrigeration ton (RT) as

$$Q_{\text{eva}} = m_{\text{ref}}\left(h_{\text{ref},2} - h_{\text{ref},1}\right) / 3.517 \tag{2.1}$$

where Q_{eva} is the heat energy absorption rate by the refrigerant in the evaporator or the cooling effect produced by the chiller in RT, m_{ref} is the mass flow rate of refrigerant in kg/s, $h_{\text{ref},1}$ is the enthalpy of refrigerant at the inlet of the evaporator at Stage 1 in kJ/kg, and $h_{\text{ref},2}$ is the enthalpy of refrigerant at the outlet of the evaporator at Stage 2 in kJ/kg. To avoid the inaccuracy of the mass flow rate measurements of two-phase (mixture of vapor and liquid) refrigerant, the cooling effect produced by a chiller is calculated by measuring the flow rate and temperature of chilled water as

$$Q_{\text{eva}} = V_{\text{chw}}\rho_{\text{chw}}c_{p,\text{chw}}\left(T_{\text{chw,in}} - T_{\text{chw,out}}\right) / 3.517 \tag{2.2}$$

where Q_{eva} is the cooling effect produced by the chiller in RT, V_{chw} is the volume flow rate of chilled water in the evaporator coils in m^3/s, ρ_{chw} is the density of chilled water in kg/m^3, $c_{p,\text{chw}}$ is the specific heat capacity of chilled water in kJ/(kg K), $T_{\text{chw,in}}$ is the temperature of chilled water at the inlet of the evaporator in °C, and $T_{\text{chw,out}}$ is the temperature of chilled water at the outlet of the evaporator in °C. The rate of heat rejection in the condenser can be calculated as

$$Q_{\text{con}} = m_{\text{ref}}\left(h_{\text{ref},3} - h_{\text{ref},4}\right) / 3.517 \tag{2.3}$$

where Q_{con} is the heat energy absorption rate by the refrigerant in the condenser or the heat energy rejection rate by the chiller in RT, m_{ref} is the mass flow rate of refrigerant in kg/s, $h_{\text{ref},3}$ is the enthalpy of refrigerant at the inlet of the condenser at Stage 3 in kJ/kg, and $h_{\text{ref},4}$ is the enthalpy of refrigerant at the outlet of the condenser at Stage 4 in kJ/kg. The heat energy rejection rate by a chiller can also be calculated as

$$Q_{\text{con}} = V_{\text{cw}}\rho_{\text{cw}}c_{p,\text{cw}}\left(T_{\text{cw,out}} - T_{\text{chw,in}}\right)/3.517 \tag{2.4}$$

where Q_{con} is the heat energy rejection rate by the chiller in RT, V_{cw} is the volume flow rate of condenser water in the condenser coils in m³/s, ρ_{cw} is the density of condenser water in kg/m³, $c_{p,\text{cw}}$ is the specific heat capacity of condenser water in kJ/(kg K), $T_{\text{cw,in}}$ is the temperature of condenser water at the inlet of the condenser in °C, and $T_{\text{cw,out}}$ is the temperature of condenser water at the outlet of the condenser in °C. The electric energy input to the compressor of the chiller can be determined as

$$W_{\text{com}} = m_{\text{ref}}\left(h_{\text{ref},3} - h_{\text{ref},2}\right)/3.517 \tag{2.5}$$

where W_{com} is the electric energy input to the compressor in RT, m_{ref} is the mass flow rate of refrigerant in kg/s, $h_{\text{ref},2}$ is the enthalpy of refrigerant at the inlet of the compressor at Stage 2 in kJ/kg, and $h_{\text{ref},3}$ is the enthalpy of refrigerant at the outlet of the compressor at Stage 3 in kJ/kg. The electric energy input to the motor of the compressor of a chiller is expressed as

$$W_{\text{chiller}} = m_{\text{ref}}\left(h_{\text{ref},3} - h_{\text{ref},2}\right)/\left(\eta_{\text{gear}}\eta_{\text{motor}}\right) \tag{2.6}$$

where W_{chiller} is the electric energy input to the motor of the compressor of the chiller in kW, η_{gear} is the efficiency of the gearbox, and η_{motor} is the efficiency of the motor of the compressor. The input energy to the motor of a compressor can be measured using a power meter. The compressor shown in Figure 2.2 is hermetically sealed. The electric energy input to the motor of a hermetically sealed compressor eventually transfers to the refrigerant. On the other hand, a fraction of the electric energy input to the motor of an open compressor is transferred to the ambient, and the remaining electric energy is transferred to the refrigerant. The actual electric energy transfer to the refrigerant should be considered during the energy balance analysis of a chiller to evaluate the accuracy of the measuring sensors. Neglecting the heat losses by convection and radiation from the exposed surfaces of the condenser, evaporator, and the casing of the compressor and motor, the equation of energy conservation for a chiller with hermetically sealed compressors gives

$$Q_{\text{eva}} + W_{\text{chiller}}/3.517 = Q_{\text{con}} \tag{2.7}$$

The equation of energy conservation for a chiller with open-type compressors is expressed as

$$Q_{\text{eva}} + W_{\text{com}} = Q_{\text{con}} \tag{2.8}$$

Equations (2.7) and (2.8) are used to evaluate the accuracy of the power meters, flow meters, and temperature sensors used in the chiller system. The efficiency of a chiller is expressed as

$$\eta_{\text{chiller}} = W_{\text{chiller}} / Q_{\text{eva}} \tag{2.9}$$

where η_{chiller} is the efficiency of the chiller in kW/RT. In fact, η_{chiller} represents the specific energy consumption of a chiller, which is defined as the electric energy consumed by a chiller in kW to produce the cooling effect in RT. The efficiency of a chiller plant is defined as

$$\eta_{\text{plant}} = \frac{W_{\text{chiller}} + W_{\text{chwp}} + W_{\text{cwp}} + W_{\text{CT}}}{Q_{\text{eva}}} \tag{2.10}$$

where η_{plant} is the efficiency of the chiller plant in kW/RT, W_{chwp} is the electric energy input to the motor of the chilled water pumps in kW, W_{cwp} is the electric energy input to the motor of the condenser water pumps in kW, and W_{CT} is the electric energy input to the motor of the fans of the cooling towers in kW. As η_{chiller}, the η_{plant} represents the specific energy consumption of a chiller plant, which is defined as the total electric energy consumed by the chillers, pumps, and cooling towers in kW to produce the cooling effect in RT. If a chiller plant contains secondary and tertiary chilled water pumps, the electric energy consumptions of the secondary and tertiary chilled water pumps are also included to calculate the chiller plant efficiency.

In most facilities, multiple chillers along with their associated pumps and cooling towers are installed within a facility to meet the cooling requirements of the various spaces. Two common designs for chilled water pumping systems are (a) primary chilled water pumping system and (b) primary-secondary chilled water pumping system. Illustrated in Figure 2.5 is a central air-conditioning plant equipped with three sets of chillers utilizing

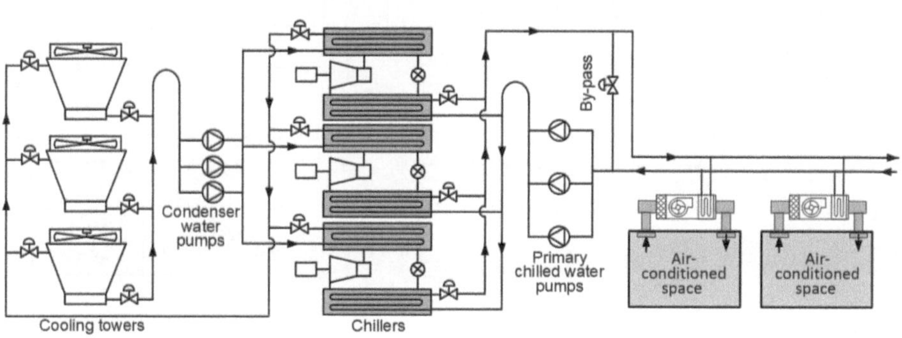

FIGURE 2.5
Chiller plant with primary chilled water pumps installed on the chilled water return header pipe.

a primary chilled water pumping system. Each set includes chilled water pumps, condenser water pumps, and cooling towers interconnected with the chillers through common header pipes. The primary chilled water pumps are typically operated at a constant speed, regardless of the cooling load in the spaces. Additionally, the speed of these primary chilled water pumps can be adjusted using variable-speed drive (VSD) to align with real-time demand for chilled water.

When the primary chilled water pumps are operated at a constant speed, the flow-modulating valve on the bypass pipe is partially opened to redirect excess chilled water when the demand from the AHUs decreases due to lower cooling loads in the spaces. Conversely, when the speeds of the primary chilled water pumps are regulated according to real-time chilled water demand, the modulating valve on the bypass pipe remains closed until the demand for chilled water of the AHUs drops below the minimum required flow for the chillers. If the demand for chilled water from the AHUs falls below this minimum requirement, the modulating valve on the bypass pipe is partially opened to compensate for the shortfall in the chilled water supply to the chillers. The reduction in chilled water pump speed during chiller part-load operation significantly decreases chilled water pump power consumption. The primary chilled water pumps could be installed on the chilled water supply header pipe instead of on the return header pipe as illustrated in Figure 2.6. Installation of the primary pumps on the supply header pipe could be an effective solution to reduce the static pressure head for the heat exchanger tubes of evaporators.

Figure 2.7 illustrates a chiller plant featuring three sets of chillers equipped with a primary-secondary chilled water pumping system. The primary chilled water pumps are typically small and operate at a constant speed, regardless of the chiller cooling load. Their role is to circulate chilled water through the chillers and a small decoupler piping loop, usually installed in

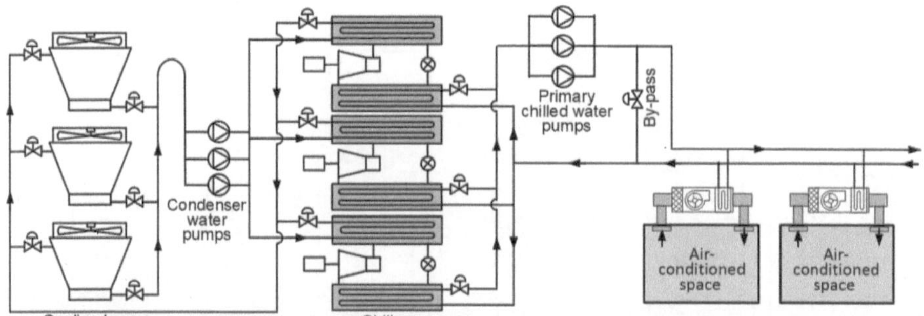

FIGURE 2.6
Chiller plant with primary chilled water pumps installed on the chilled water supply header pipe.

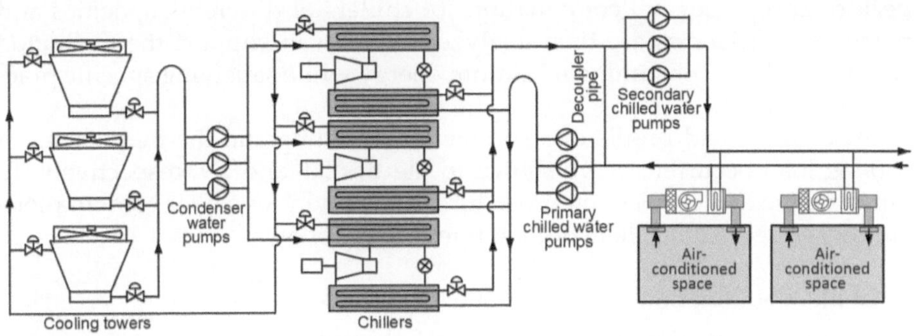

FIGURE 2.7
Chiller plant with primary and secondary chilled water pumping system.

the chiller plant room. Contrarily, the secondary chilled water pumps are larger and designed to supply chilled water to all the AHUs within the building. Their speed is modulated using VSDs based on real-time demand for chilled water, resulting in substantial energy savings for the secondary chilled water pumps, particularly at low part-load conditions. If the flow rate of the primary chilled water pumps exceeds that of the secondary chilled water pumps, chilled water flows from the supply header pipe to the return header pipe through the decoupler pipe. In such cases, one chiller, together with its associated pumps and cooling towers, is turned off if the chilled water flow through the decoupler pipe matches or exceeds the flow of one chiller. Conversely, if the flow rate of chilled water through the secondary chilled water pumps becomes higher than that of the primary chilled water pumps, chilled water starts to flow from the return header to the supply header through the decoupler pipe. In this scenario, another chiller, together with its associated pumps and cooling towers, is turned on.

2.2 Chiller Sizing and Configuration

The efficiency of a chiller depends on its loading. The optimum loading of a chiller depends on several factors such as type of compressor, cooling capacity, type of refrigerant, design of heat exchangers of the evaporator and condenser, and control strategies. To operate a chiller plant in an energy-efficient manner, the chillers should be operated at the region of optimum loadings while supporting the dynamic cooling loads of the building. The cooling load profile of existing buildings is determined by trend logging the return temperature, supply temperature, and flow rate of chilled water through the evaporators of the operating chillers. For new buildings, the cooling load profile is determined using energy simulation software. The energy performance of several combinations of chillers of different capacities and part-load efficiencies are then analyzed which can support the evaluated cooling loads by consuming minimum energy while satisfying specific practical constraints.

The cooling load profile of a commercial building and the frequencies of cooling load occurrences are shown in Figures 2.8 and 2.9, respectively. To illustrate the chiller selection procedure, two options of chiller combinations are considered. The chiller combinations are as follows:

Chiller combination-a: 3 nos. of 400 RT chillers

Chiller combination-b: 2 nos. of 600 RT chillers

Part-load efficiencies for the chillers are summarized in Table 2.1. Chiller manufacturers provide part-load efficiencies of chillers of different capacities under different operating conditions.

Daily electrical energy consumption for chiller combination-a (3 nos. of 400 RT chillers) based on an equal percentage of loading of each operating chiller can be calculated as elaborated in Table 2.2.

Average daily energy consumption for chiller combination-a = 7197.4 kWh/day.

Assuming that the chiller plant is operated for 365 days a year, the annual energy consumption = 7197.4 × 365 = 2,627,033 kWh/year.

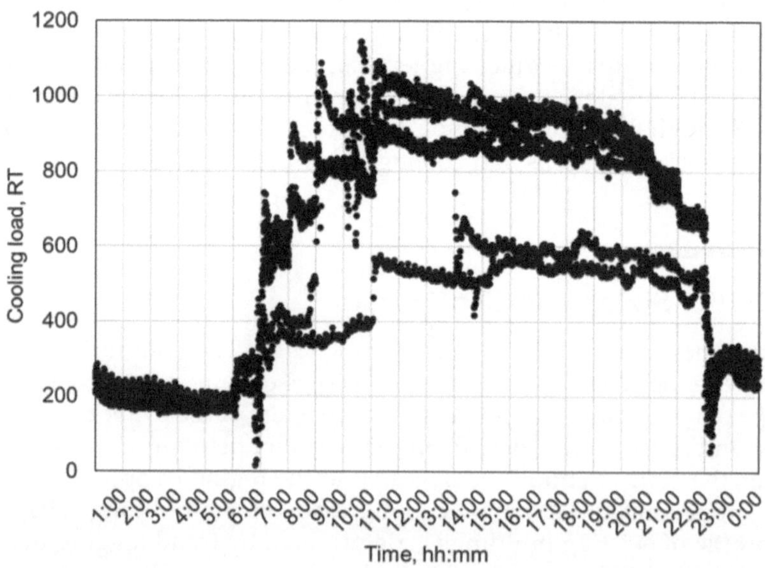

FIGURE 2.8
Cooling load profile of a commercial building.

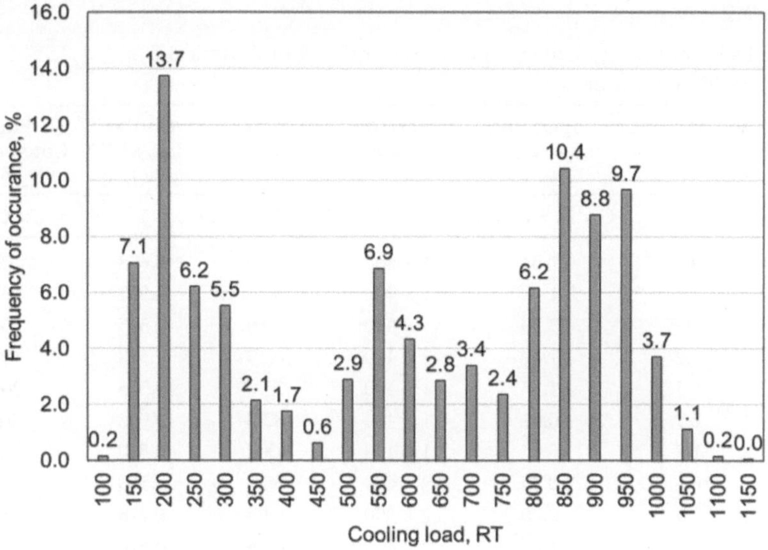

FIGURE 2.9
Frequencies of cooling load occurrences.

TABLE 2.1
Part-Load Efficiencies of Chillers

Chiller Loading, %	400 RT	600 RT
100	0.53	0.54
90	0.51	0.52
80	0.501	0.5
70	0.508	0.49
60	0.521	0.48
50	0.545	0.49
40	0.579	0.5
30	0.632	0.55
20	0.724	0.66
10	0.80	0.77

Similarly, daily electrical energy consumption for chiller combination-b (2 nos. of 600 RT chillers) based on an equal percentage of loading of each operating chiller can be calculated as illustrated in Table 2.3.

Average daily energy consumption for chiller combination-b = 7054.4 kWh/day.

Assuming that the chiller plant is operated for 365 days a year, the annual energy consumption = 7054.4 × 365 = 2,574,873 kWh/year.

TABLE 2.2

Daily Electrical Energy Consumption for Chiller Combination-a (3 Nos. of 400 RT Chillers)

Av. RT	Occurrence, %	Operating Hours/Day	Operating Chiller	Part Load, %	Chiller Eff, kW/RT	Energy Consumption, kWh
A	B	C = 24 × (B/100)	D	E = 100 × A/D	F	A × C × F
100	0.2	0.04	1 × 400	25.0	0.678	2.6
150	7.1	1.69	1 × 400	37.5	0.592	150.4
200	13.7	3.30	1 × 400	50.0	0.545	359.3
250	6.2	1.49	1 × 400	62.5	0.518	193.0
300	5.5	1.33	1 × 400	75.0	0.505	200.8
350	2.1	0.51	1 × 400	87.5	0.508	91.4
400	1.7	0.42	1 × 400	100.0	0.530	88.9
450	0.6	0.15	2 × 400	56.3	0.530	35.2
500	2.9	0.70	2 × 400	62.5	0.518	180.0
550	6.9	1.65	2 × 400	68.8	0.510	461.9
600	4.3	1.04	2 × 400	75.0	0.505	314.3
650	2.8	0.68	2 × 400	81.3	0.502	223.1
700	3.4	0.82	2 × 400	87.5	0.508	290.3
750	2.4	0.56	2 × 400	93.8	0.518	219.1
800	6.2	1.48	2 × 400	100.0	0.530	627.0
850	10.4	2.51	3 × 400	70.8	0.507	1080.5
900	8.8	2.11	3 × 400	75.0	0.505	955.9
950	9.7	2.32	3 × 400	79.2	0.502	1105.3
1000	3.7	0.89	3 × 400	83.3	0.504	447.7
1050	1.1	0.27	3 × 400	87.5	0.508	142.2
1100	0.2	0.04	3 × 400	91.7	0.513	21.5
1150	0.05	0.01	3 × 400	95.8	0.522	7.1
			Daily energy consumption, kWh/day = 7197.4			

The annual electric energy consumption of chiller combination-b is lower than that of chiller combination-a, suggesting chiller combination-b is a potentially superior option. The energy consumption for other chiller combinations could be calculated following the procedures outlined in Table 2.2 and find the most energy-efficient chiller combination. Chillers run at a higher percentage of loading if the capacity of the chillers is small. The energy efficiency of larger chillers is usually better. Consequently, the energy consumption of chiller combination-b, comprising two 600 RT chillers, is lower. It is a common practice to install at least one standby chiller in the plant room of commercial buildings to handle the cooling load during the period of maintenance or chiller breakdowns. For combination-b,

TABLE 2.3
Daily Electrical Energy Consumption for Chiller Combination-b (2 nos. of 600 RT Chillers)

Av. RT	Occurrence, %	Operating Hours/Day	Operating Chiller	Part Load, %	Chiller Eff, kW/RT	Energy Consumption, kWh
A	B	C = 24 × (B/100)	D	E = 100 × A/D	F	A × C × F
100	0.2	0.04	1 × 600	16.7	0.697	2.7
150	7.1	1.69	1 × 600	25.0	0.605	153.7
200	13.7	3.30	1 × 600	33.3	0.533	351.6
250	6.2	1.49	1 × 600	41.7	0.498	185.7
300	5.5	1.33	1 × 600	50.0	0.490	195.0
350	2.1	0.51	1 × 600	58.3	0.482	86.7
400	1.7	0.42	1 × 600	66.7	0.487	81.6
450	0.6	0.15	1 × 600	75.0	0.495	32.9
500	2.9	0.70	1 × 600	83.3	0.507	176.2
550	6.9	1.65	1 × 600	91.7	0.523	474.3
600	4.3	1.04	1 × 600	100.0	0.540	336.4
650	2.8	0.68	2 × 600	54.2	0.486	215.8
700	3.4	0.82	2 × 600	58.3	0.482	275.4
750	2.4	0.56	2 × 600	62.5	0.483	204.2
800	6.2	1.48	2 × 600	66.7	0.487	575.8
850	10.4	2.51	2 × 600	70.8	0.491	1045.2
900	8.8	2.11	2 × 600	75.0	0.495	937.9
950	9.7	2.32	2 × 600	79.2	0.499	1099.9
1000	3.7	0.89	2 × 600	83.3	0.507	450.1
1050	1.1	0.27	2 × 600	87.5	0.515	144.2
1100	0.2	0.04	2 × 600	91.7	0.523	21.9
1150	0.05	0.01	2 × 600	95.8	0.532	7.3
			Daily energy consumption, kWh/day = 7054.4			

a standby chiller with a capacity matching that of the main chillers (600 RT) is required, bringing the total installed capacity to 1,800 RT. Conversely, combination-a requires a standby chiller with a capacity of 400 RT, resulting in a total installed capacity of 1600 RT. Consequently, the installation cost for combination-a is lower. Despite the lower annual energy savings of combination-a compared to combination-b (2,627,033 – 2,574,873 = 52,160 kWh/year), combination-a still presents a viable option for implementation. However, combination-a requires more plant room space due to the installation of four chillers, whereas combination-b only requires three. Other practical constraints should also be considered during the selection of chillers.

2.3 Chiller Sequencing and Control Strategy

The main objective of chiller sequencing is to select different combinations of chillers to be operated at specific loadings to support different cooling loads of spaces in an energy-efficient manner. The selected combinations of chillers should be able to supply the required amount of chilled water at the pre-set temperature to each AHU and provide the required sensible and latent cooling. The typical chiller sequencing strategy is illustrated in Figure 2.10.

The typical sequencing strategy distributes an equal percentage of cooling load to each operating chiller. However, energy consumption can be optimized by allocating the cooling load based on each chiller's optimum loading, corresponding to its best part-load efficiency. This optimization approach should also account for the impacts of variable chilled and condenser water flow rates and temperatures on energy consumption. Additionally, the performance characteristics of chilled water pumps, condenser water pumps, cooling towers, AHU fans, and the temperature, relative humidity, and fresh air requirements in air-conditioned spaces must be considered. This comprehensive approach will enable the development of a holistic control strategy to operate the entire chiller plant with maximum energy efficiency.

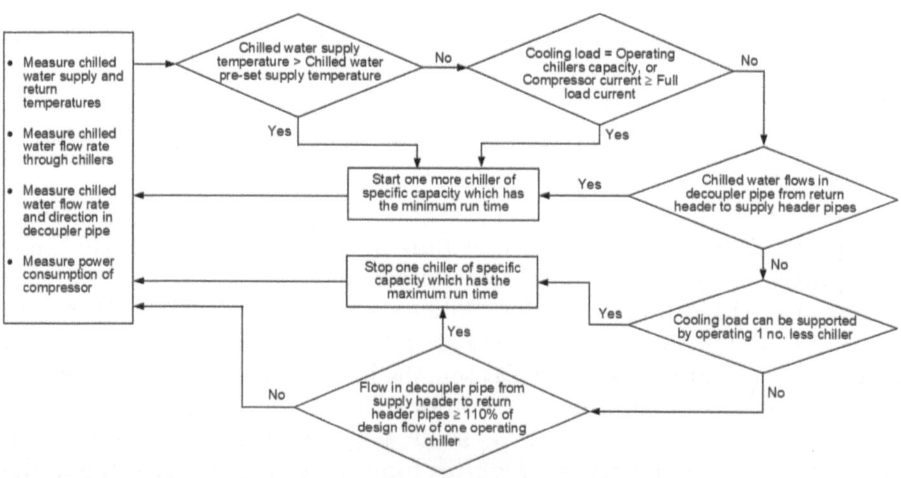

FIGURE 2.10
Typical strategy for turning on a chiller.

2.4 Control Strategies for Pumps to Regulate Water Flow to Heat Exchangers

The static pressure head in chilled water loops designed to circulate water to AHUs in high-rise buildings, or a mix of high-rise and low-rise buildings, is typically very high. To withstand this pressure, the chilled water loops along with the evaporator heat exchangers of chillers and the cooling coils of AHUs are constructed with thicker-walled pipes and tubes. Pumps, valves, and associated sensors are selected to match this high static pressure, leading to significantly increased capital and maintenance costs for the chiller system. To address this issue, flat plate heat exchangers can be strategically installed to isolate high static pressure areas and equalize the pressure between the primary and secondary sides of the chiller plants as illustrated in Figure 2.11. Two conventional control strategies are commonly employed to modulate the speed of chilled water pumps on the primary side of the heat exchangers. The primary goal of these strategies is to use VSDs to regulate pump speed and ensure that the appropriate amount of chilled water circulates through the heat exchangers, maintaining the pre-set chilled water supply temperature on the secondary side.

The first conventional control strategy, known as a pressure-based control strategy, regulates the speed of the secondary chilled water pumps installed on the primary side of heat exchangers to maintain a pre-set differential pressure between the supply and return header pipes on the primary side, as illustrated in Figure 2.12 [3]. Temperature sensors installed on the chilled water supply pipes after the heat exchangers (i.e. on the secondary side of the heat exchanger) measure the chilled water supply temperature. This measured temperature is then used as an input for the temperature controller, which compares it to the set-point temperature and adjusts the opening of the flow-modulating valves on the primary side chilled water return pipes accordingly. If the measured temperature is higher than the set-point due to increased cooling demand from the AHUs, the temperature controller will open the flow-modulating valve wider. Consequently, the measured differential pressure between the primary side chilled water supply and return header pipes will drop below the set-point differential pressure. The differential pressure controller will then increase the speed of the secondary chilled water pumps located on the primary side of the heat exchanger by adjusting the VSD frequency to raise the differential pressure back to the set-point.

The second conventional control strategy, known as the flow-based control strategy, modulates the speed of the secondary chilled water pumps installed on the primary side of heat exchangers to maintain a pre-set chilled water temperature of the chilled water supply header pipes located on the secondary side of the heat exchanger, as depicted in Figure 2.13 [3]. This strategy

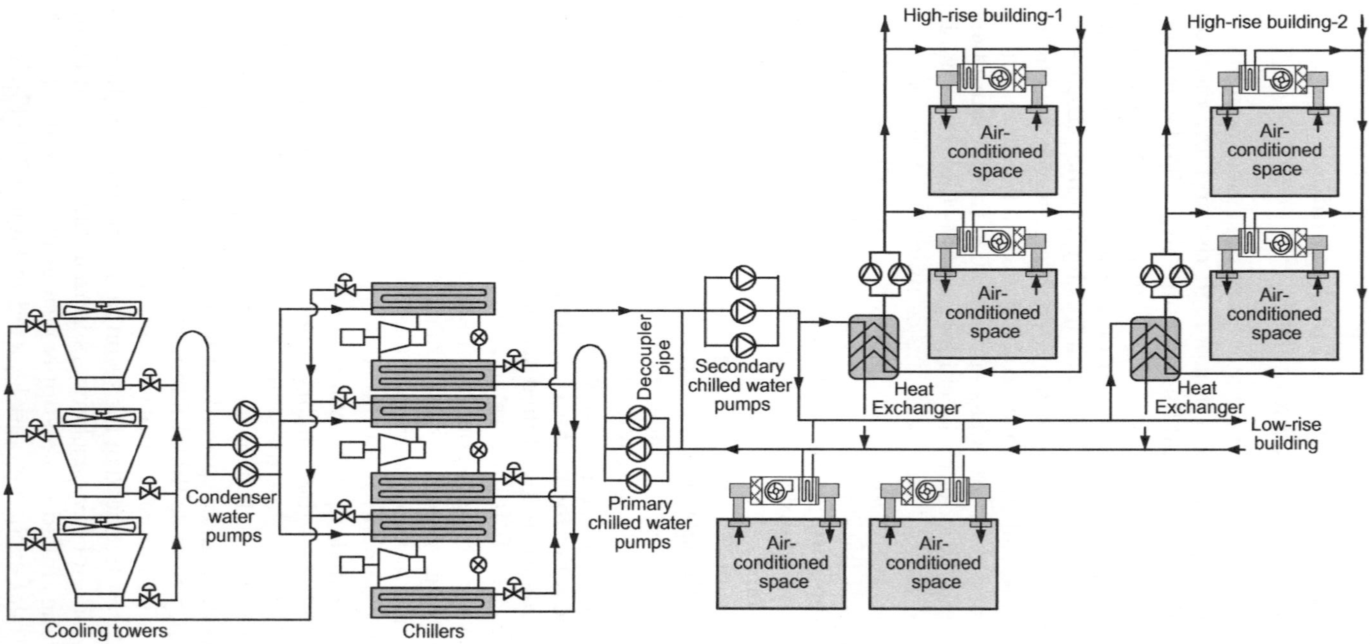

FIGURE 2.11
Flat plate heat exchangers are installed to isolate high static pressure heads of high-rise building 1 and 2.

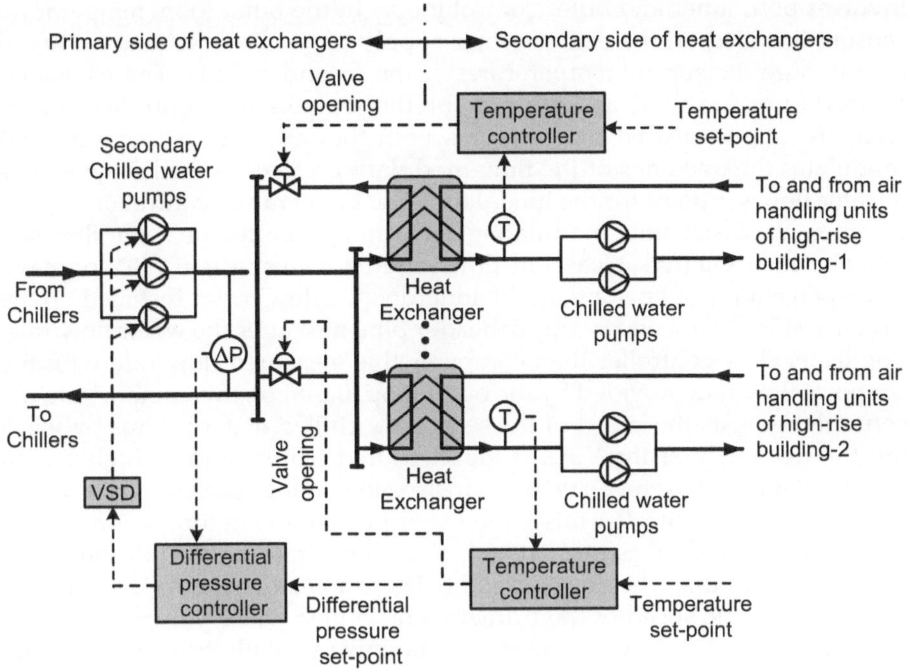

FIGURE 2.12
Pressure-based control strategy to regulate the speed of secondary chilled water pumps [3].

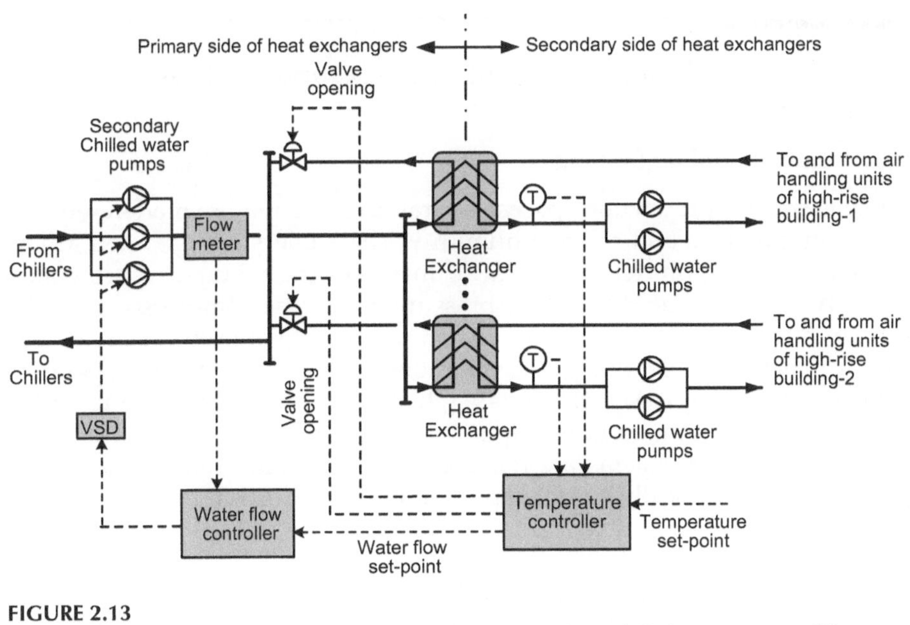

FIGURE 2.13
Flow-based control strategy to regulate the speed of secondary chilled water pumps [3].

involves both inner and outer control loops. In the outer loop, temperature sensors installed on the chilled water supply pipes after the heat exchangers measure the supply temperatures at the secondary side. The measured temperatures are used as an input for the temperature controller, which compares the measured temperatures with the set-point temperature and modulates the openings of the flow-modulating valves as well as generates a water flow set-point for the inner loop. The temperature controller records the opening of each flow-modulating valve and generates the water flow set-point by ensuring that at least one flow-modulating valve is at 100% or maximum percentage of opening. In the inner loop, a flow meter installed on the primary side chilled water supply header pipe measures the water flow rate. The water flow controller then compares this measured flow rate with the set-point flow rate provided by the outer loop. Based on this comparison, the controller adjusts the speed of the secondary chilled water pumps by varying the frequency of the VSD. If the measured temperature is higher than the set-point temperature due to an increased cooling load from the AHUs, the temperature controller raises the water flow set-point for the inner loop. Consequently, the measured chilled water flow rate on the primary side becomes lower than the new set-point. The water flow controller responds by increasing the speed of the primary side chilled water pumps, adjusting the VSD frequency to match the measured flow rate with the new set-point.

2.5 Energy Performance Enhancement Opportunities of Chiller

2.5.1 Cascade Chiller Systems

Some industrial production spaces, laboratories, etc. require moderately low space temperature and relative humidity. The chillers are operated at low chilled water supply temperatures to maintain the required space conditions. As a result, the differential pressure across the compressor becomes too large for a single compressor system to operate efficiently. The energy performance of chillers for such situations can be enhanced by operating two or more compressors in series, which is known as a cascade chiller system [4]. Even though the complexity, capital cost, and maintenance cost for cascade chiller systems are relatively high, cascade chiller systems still become an economically feasible option for large industrial applications due to their better energy efficiency. A two-stage cascade chiller system is shown in Figure 2.14.

The two vapor compression refrigeration cycles are connected through a heat exchanger, which serves as the evaporator for the vapor compression

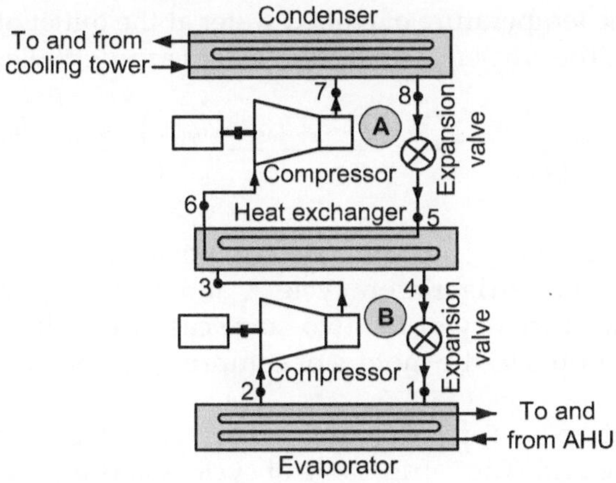

FIGURE 2.14
Two-stage cascade chiller system.

refrigeration cycle A and the condenser for the vapor compression refrigeration cycle B. Assuming the negligible heat loss from the surface of the heat exchanger, the heat transfer from the refrigerant of cycle B should be equal to the heat transfer rate to the refrigerant of cycle A:

$$m_{\text{ref},A}\left(h_{\text{ref},6}-h_{\text{ref},5}\right)=m_{\text{ref},B}\left(h_{\text{ref},3}-h_{\text{ref},4}\right) \tag{2.11}$$

where $m_{\text{ref,A}}$ is the mass flow rate of refrigerant in cycle A in kg/s, $m_{\text{ref,B}}$ is the mass flow rate of refrigerant in cycle B in kg/s, $h_{\text{ref,3}}$ is the enthalpy of refrigerant of cycle A at the inlet of the heat exchanger at Stage 3 in kJ/kg, $h_{\text{ref,4}}$ is the enthalpy of refrigerant of cycle A at the outlet of the heat exchanger at Stage 4 in kJ/kg, $h_{\text{ref,5}}$ is the enthalpy of refrigerant of cycle B at the inlet of the heat exchanger at Stage 5 in kJ/kg, and $h_{\text{ref,6}}$ is the enthalpy of refrigerant of cycle B at the outlet of the heat exchanger at Stage 6. The cooling effect produced by the two-stage cascade chiller can be calculated by measuring the flow rate and temperatures of chilled water or using the enthalpies of refrigerant as

$$Q_{\text{eva},B}=\frac{V_{\text{chw}}\rho_{\text{chw}}c_{\text{p,chw}}\left(T_{\text{chw,in}}-T_{\text{chw,out}}\right)}{3.517}=\frac{m_{\text{ref},B}\left(h_{\text{ref},2}-h_{\text{ref},1}\right)}{3.517} \tag{2.12}$$

where $Q_{\text{eva,B}}$ is the cooling effect produced by the two-stage cascade chiller in RT, V_{chw} is the volume flow rate of chilled water in the evaporator coils of cycle B in m^3/s, ρ_{chw} is the density of chilled water of cycle B in kg/m^3, $c_{\text{p,chw}}$ is the specific heat capacity of chilled water of cycle B in kJ/(kg K), $T_{\text{chw,in}}$ is the temperature of chilled water at the inlet of the evaporator of cycle B in °C, and $T_{chw,out}$ is the temperature of chilled water at the outlet of the evaporator of cycle B in °C. The efficiency of the two-stage cascade chiller is expressed as

$$\eta_{\text{chiller,cascade}} = \frac{W_{\text{chiller},A} + W_{\text{chiller},B}}{Q_{\text{eva},B}} = \frac{m_{\text{ref},A}\left(h_{\text{ref},7} - h_{\text{ref},6}\right) + m_{\text{ref},B}\left(h_{\text{ref},3} - h_{\text{ref},2}\right)}{Q_{\text{eva},B}} \quad (2.13)$$

where $W_{\text{chiller,A}}$ and $W_{\text{chiller,B}}$ are the electric power inputs to the motors of compressors of the refrigeration cycle A and B in kW, respectively, and $\eta_{\text{chiller,cascade}}$ is the efficiency of the two-stage cascade chiller in kW/RT. The electric power inputs to the motors of compressors can be measured using the power meter.

The temperature–entropy diagram of the two-stage cascade chiller is shown in Figure 2.15. The refrigerants of cycle A and B are not mixed in the heat exchanger. Hence, refrigerants with desirable characteristics can be used in each cycle. It is evident from Figure 2.15 that the cooling capacity increases and the compressor work decreases as a result of cascading. Therefore, cascading improves the efficiency of the chiller systems.

2.5.2 Multistage Vapor Compression Chiller Systems

The heat exchanger between the vapor compression refrigeration cycles A and B of the cascade chiller systems can be replaced by a flash chamber and a mixing box when the refrigerant used throughout the cascade chiller system is the same. The flash chamber has better heat transfer characteristics. Such systems are called multistage vapor compression chiller systems. A two-stage vapor compression chiller system is shown in Figure 2.16. The pressure and the corresponding boiling temperature of the refrigerant drop after flowing through the expansion valve of cycle A. As a result, part of the liquid refrigerant evaporates in the flush chamber by absorbing the latent heat of vaporization from the refrigerant. The temperature of the flashed saturated vapor

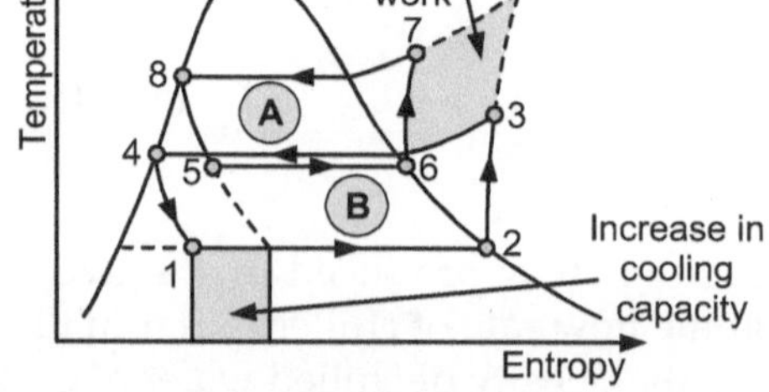

FIGURE 2.15
Temperature–entropy diagram of two-stage cascade chiller.

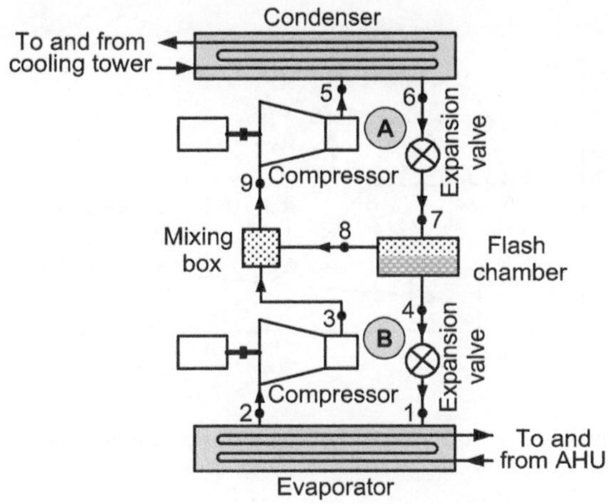

FIGURE 2.16
Two-stage vapor compression chiller system with a flash chamber.

and the remaining liquid refrigerant drops. The low-temperature saturated vapor (State 8) is mixed with the superheated vapor (State 3) from the low-pressure compressor of cycle B. The mixing chamber acts as an intercooler of the two-stage compression process, and the compressor work decreases [4]. The temperature-entropy diagram of the two-stage vapor compression chiller system is shown in Figure 2.17. The multistage vapor compression chiller systems improve the efficiency of the chiller. The cooling effect produced per unit mass of refrigerant flowing through the evaporator of cycle B is increased as saturated liquid refrigerant enters State 1 to the evaporator of cycle B. However, the cooling capacity of the chillers drops slightly due to the reduction of the refrigerant flow rate through the evaporator of cycle B. The mass flow rate of refrigerant through the evaporator of cycle B can be calculated as

$$m_{\text{ref},B} = m_{\text{ref},A}\left(1-\frac{h_{\text{ref},7}-h_{\text{ref},4}}{h_{\text{ref},8}-h_{\text{ref},4}}\right) \tag{2.14}$$

where $m_{\text{ref,A}}$ and $m_{\text{ref,B}}$ are the refrigerant flow rates through the vapor compression cycle A and B in kg/s, respectively. The cooling effect or the heat removal rate from the air-conditioned spaces can be calculated as

$$Q_{\text{eva},B} = m_{\text{ref},A}\left(1-\frac{h_{\text{ref},7}-\text{h}_{\text{ref},4}}{h_{\text{ref},8}-h_{\text{ref},4}}\right)\frac{\left(h_{\text{ref},2}-h_{\text{ref},1}\right)}{3.517} = \frac{V_{\text{chw}}\rho_{\text{chw}}c_{\text{p,chw}}\left(T_{\text{chw,in}}-T_{\text{chw,out}}\right)}{3.517} \tag{2.15}$$

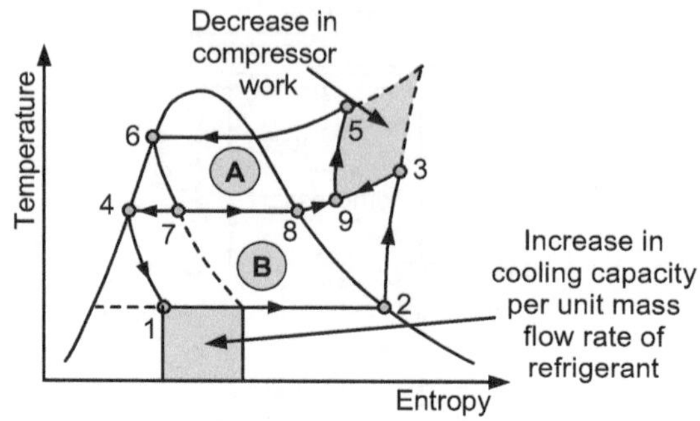

FIGURE 2.17
Temperature–entropy diagram of the two-stage vapor compression chiller.

where $Q_{eva,B}$ is the cooling effect produced by the two-stage vapor compression chiller in RT, V_{chw} is the volume flow rate of chilled water in the evaporator coils of cycle B in m³/s, ρ_{chw} is the density of chilled water of cycle B in kg/m³, $c_{p,chw}$ is the specific heat capacity of chilled water of cycle B in kJ/(kg K), $T_{chw,in}$ is the temperature of chilled water at the inlet of the evaporator of cycle B in °C, and $T_{chw,out}$ is the temperature of chilled water at the outlet of the evaporator of cycle B in °C. The efficiency of the two-stage vapor compression chiller is determined as

$$\eta_{\text{chiller,two-stage}} = \frac{W_{\text{chiller},A} + W_{\text{chiller},B}}{Q_{\text{eva},B}} = \frac{m_{\text{ref},A}\left(h_{\text{ref},5} - h_{\text{ref},9}\right) + m_{\text{ref},B}\left(h_{\text{ref},3} - h_{\text{ref},2}\right)}{Q_{\text{eva},B}} \tag{2.16}$$

where $W_{chiller,A}$ and $W_{chiller,B}$ are the electric power inputs to the motors of compressors of the refrigeration cycles A and B in kW, respectively, and $\eta_{chiller,two\text{-}stage}$ is the efficiency of the two-stage vapor compression chiller in kW/RT. The electric power inputs to the motors of compressors can be measured using the power meter. The efficiency of a two-stage cascade chiller or a two-stage vapor compression chiller improves considerably relative to the single-stage vapor compression chiller. Under similar operating conditions, the efficiency of the two-stage vapor compression chillers improves slightly compared to the two-stage cascade chillers. However, the cooling capacity of the two-stage vapor compression chillers drops slightly compared to the two-stage cascade chillers. The capital cost and the maintenance cost of the two-stage vapor compression chillers are lower than the two-stage cascade chillers as the heat exchanger of the cascade chillers is replaced by the flash chamber and the mixing box in the multistage vapor compression chillers.

2.5.3 Optimization of Sequencing and Loading of Chillers

The dynamic cooling loads of air-conditioned spaces are usually supported by operating several chillers of different cooling capacities and part-load efficiencies installed in the chiller plant rooms. A selected number of chillers are commonly operated at an equal percentage of loadings to support the cooling loads. Suppose the maximum cooling load of a building is 900 RT which is supported by three chillers of cooling capacities 300, 350, and 400 RT. The part-load efficiencies of the chillers are shown in Table 2.4. The average dynamic cooling loads of the building and the sequencing of chillers are summarized in Table 2.5.

The power consumption of the chillers at different loadings can be calculated using the part-load efficiencies as presented in Table 2.6.

The equations of power consumptions of the chillers at different part loads can be developed by fitting the data of the part loads (Column A of Table 2.6) and the corresponding power consumptions of chillers (Columns E, F, and G of Table 2.6) as follows:

$$W_{300RT} = -191.52(PLR_{300RT})^3 + 729.74(PLR_{300RT})^2 - 535.6(PLR_{300RT}) + 197.47 \quad (2.17)$$

TABLE 2.4

Part-Load Efficiencies of Chillers

Part Load	300 RT	350 RT	400 RT
0.3	1.081	1.012	1.09
0.4	0.731	0.844	0.858
0.5	0.588	0.721	0.718
0.6	0.541	0.634	0.627
0.7	0.545	0.582	0.569
0.8	0.575	0.561	0.535
0.9	0.618	0.572	0.522
1	0.667	0.613	0.525

TABLE 2.5

Building Average Cooling Load and Present Sequencing of Chillers

Cooling Load, RT	Operating Hours, hr	Present Operating Chillers
900	2	400 + 350 + 300
800	3	400 + 350 + 300
700	2.5	400 + 350 + 300
600	2	400 + 350
500	1.8	400 + 350
400	1.6	350 + 300

TABLE 2.6
Power Consumption of Chillers at Different Loadings

Part Load	Part-Load Efficiency, kW/RT			Power Consumption of Chillers, kW		
	300 RT	350 RT	400 RT	300 RT	350 RT	400 RT
A	B	C	D	E = A × B × 300	F = A × C × 350	G = A × D × 400
0.3	1.081	1.012	1.09	97.29	106.26	130.8
0.4	0.731	0.844	0.858	87.72	118.16	137.28
0.5	0.588	0.721	0.718	88.2	126.175	143.6
0.6	0.541	0.634	0.627	97.38	133.14	150.48
0.7	0.545	0.582	0.569	114.45	142.59	159.32
0.8	0.575	0.561	0.535	138	157.08	171.2
0.9	0.618	0.572	0.522	166.86	180.18	187.92
1	0.667	0.613	0.525	200.1	214.55	210

$$W_{350RT} = 513.24(PLR_{350RT})^3 - 812.97(PLR_{350RT})^2 + 498.17(PLR_{350RT}) + 16.123 \tag{2.18}$$

$$W_{400RT} = 205.76(PLR_{400RT})^3 - 271.82(PLR_{400RT})^2 + 180.59(PLR_{400RT}) + 95.486 \tag{2.19}$$

The power consumption of each chiller and the daily energy consumption of the operating chillers can be calculated as illustrated in Table 2.7.

The average daily energy consumption of the operating three chillers based on an equal percentage of loadings for the chillers = 4912.2 kWh/day.

Annual electric energy consumption based on conventional control strategies of equal percentage of loadings for the chillers = 4912.2 × 365 = **1,792,952 kWh/year**.

Average daily cooling load = 900 × 2 +800 × 3+700 × 2.5 + 600 × 2 +500 × 1.8 + 400 × 1.6 = 8,690 RTh/day.

The present weighted average efficiency of chillers = 4912.2/8690 = **0.5653 kW/RT**.

The energy consumption of the chillers can be minimized by operating the chillers at different optimized percentages of loadings instead of a constant percentage of loading to support the different dynamic cooling loads of the building. Genetic algorithm (GA), particle swarm optimization algorithm,

TABLE 2.7

Daily Energy Consumption of Chillers Based on an Equal Percentage of Loadings for the Operating Chillers

Cooling Load, RT	Operating Hours, hr	Present Operating Chillers	Chiller Loading	Power Consumption of Chillers, kW			Total Energy, kWh/day
				300 RT, kW	350 RT, kW	400 RT, kW	
A	B	C	D = A/C	E = Eq. (2.17)	F = Eq. (2.18)	G = Eq. (2.19)	H = B (E + F + G)
900	2	400 + 350 + 300	0.857	153.9	169.0	180.1	1006.2
800	3	400 + 350 + 300	0.762	128.3	150.8	166.3	1336.0
700	2.5	400 + 350 + 300	0.667	108.0	139.0	156.0	1007.5
600	2	400 + 350	0.800	0.0	157.1	171.3	657.0
500	1.8	400 + 350	0.667	0.0	139.0	156.0	531.0
400	1.6	350 + 300	0.615	99.6	134.4	0.0	374.4
							Total = 4912.2

linear programming, Greedy algorithm, etc. can be used to determine the optimized percentage of loadings for the chillers. The GA has been recognized as an effective tool for the optimization of problems with linear and nonlinear constraints, equality, and inequality constraints. The syntax for finding the global minimum of an objective function using the GA:

$$[X,Z] = \text{ga}(\text{objfcn}, \text{nvars}, A, b, \text{Aeq}, \text{beq}, \text{LB}, \text{UB}, \text{nonlcon}, \text{options}) \quad (2.20)$$

where

objfcn = Objective function

nvars = Number of variables

A = Coefficient matrix of variables of inequality linear constraints

b = Matrix of constants of inequality linear constraints

Aeq = Coefficient matrix of variables of equality linear constraints

beq = Matrix of constants of equality linear constraints

LB = Lower bound for variables

UB = Upper bound for variables

nonlcon = Nonlinear constraints

options = Options for setting population size, tolerance for function, and tolerance for constraints.

The optimized loadings of the chillers to support the dynamic cooling loads of the building determined using the GA and the daily energy consumption of the chillers are summarized in Table 2.8.

Average daily energy consumption of the operating chillers based on optimized loadings for the chillers = 4698.7 kWh/day.

Annual electric energy consumption based on optimized loadings for the chillers

= 4698.7 × 365 = **1,715,009 kWh/year.**

Annual electric energy saving due to the optimization of loading for the chillers

= 1,792,952 – 1,715,009 = **77,943 kWh/year.**

Weighted average efficiency of chillers after optimization = 4698.7/8690

= **0.5401 kW/RT**.

Percentage improvement of chiller efficiency = 100 × (0.5653 – 0.5401)/0.5653

= **4.45%**.

TABLE 2.8

Optimized Cooling Loads and Daily Energy Consumption of Chillers Using GA

Cooling Load, RT	Operating Hours, hr	Optimized Loading of Chillers Using GA			Power Consumption of Chillers, kW			Total Energy, kWh/day
		400 RT	350 RT	300 RT	400 RT, kW	350 RT, kW	300 RT, kW	
A	B	C	D	E	F = Eq. (2.17)	G = Eq. (2.18)	H = Eq. (2.19)	I = B (F + G + H)
900	2	0.9778	0.8428	0.7130	204.5	165.8	117.1	974.9
800	3	0.8623	0.7652	0.6243	181.0	151.3	100.9	1299.5
700	2.5	0.9728	0.8882	0.0000	203.4	176.9	0.0	950.6
600	2	0.8682	0.7221	0.0000	182.0	145.2	0.0	654.4
500	1.8	0.8204	0.0000	0.5727	174.3	0.0	94.1	483.2
400	1.6	1.0000	0.0000	0.0000	210.0	0.0	0.0	336.0
								Total = 4698.7

The annual energy savings and the improvement of the chiller efficiency depend on the dynamic cooling loads of the building and the part-load efficiency profiles of the installed chillers.

2.5.4 Effect of Chilled and Condenser Water Temperatures Reset

The impact of adjusting the chilled water supply temperature is illustrated in Figure 2.18. Since the saturation pressure of refrigerant within the evaporator depends on its temperature, raising the chilled water supply temperature results in an increase in refrigerant pressure inside the evaporator. The refrigeration cycles for low and increased chilled water supply temperatures are represented by 1-2-3-4 and 5-6-7-3, respectively. As illustrated in Figure 2.18, elevating the chilled water supply temperature leads to a decrease in the difference between condenser and evaporator pressures (known as compressor lift) from ($P_{cond} - P_{eva,2}$) to ($P_{cond} - P_{eva,6}$), while enhancing the cooling effect per unit mass of flowing refrigerant from ($h_2 - h_1$) to ($h_6 - h_5$). Consequently, the lift of the compressor decreases, and the refrigerant flow rate through the chiller reduces to support the cooling load of the spaces, resulting in reduced compressor power consumption and improved energy efficiency of the chiller [5]. However, the increase in chilled water supply temperature could negatively affect the AHU in delivering the required sensible and latent cooling. The AHU's ability to remove moisture depends on the temperature of the chilled water supplied to its cooling coil. If the chilled water supply temperature increases, the circulating air passing over the AHU's cooling coil may not cool sufficiently to reach the required dew point temperature. It is crucial to ensure that the cooling coil of the AHU is meticulously designed and possesses adequate surface area to meet the cooling demands associated with the higher chilled water supply temperature.

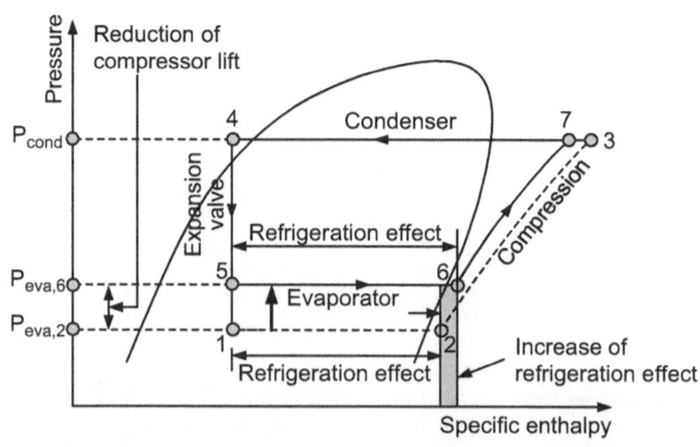

FIGURE 2.18
Effect of chilled water temperature reset.

An improvement in chiller efficiency of approximately 3% can be achieved by raising every 1°C of chilled water supply temperature. Chilled water systems are typically engineered to maintain a supply temperature of 6.7°C to handle peak cooling demands. Since chillers often operate at partial loads, supplying chilled water at 6.7°C to AHU cooling coils is not essential during these times. Thus, adjusting the chilled water supply temperature to a higher value under such conditions is both feasible and advantageous. Suppose the average cooling load of a building is 1,000 RT, which is supported by operating a chiller with a chilled water supply temperature of 6.7°C and an efficiency of 0.5 kW/RT. The building operates for 10 hours per day over 350 days in a year. Assuming a 3% improvement in chiller efficiency for every 1°C increase in chilled water supply temperature, the annual energy savings resulting from increasing the chilled water supply temperature by 8°C can be estimated as follows:

Average cooling load of the building = 1,000 RT.

Present chiller efficiency = 0.5 kW/RT.

Present chilled water supply temperature = 6.7°C.

Proposed chilled water supply temperature = 8.0°C.

Proposed increment of chilled water supply temperature = 8.0 – 6.7 = 1.3°C.

Chiller efficiency is improved by 3% for every 1°C increase in chilled water supply temperature.

Improvement of chiller efficiency due to the increase of chilled water supply temperature by 1.3°C = 1.3 × 3 = 3.9%.

Chiller new efficiency = 0.5 kW/RT × (1 – 3.9/100) = 0.4805 kW/RT.

The chiller will be operated for 10 hrs/day and 350 days/year to support the cooling load of the building.

Annual energy savings = (0.5 – 0.4805) × 1000 × 10 × 350 = 68,250 kWh/year.

Figure 2.19 illustrates the impact of resetting the condenser water supply temperature. As the temperature of the condenser water supply drops, the pressure of the refrigerant within the condenser also decreases. The refrigeration cycles for high and low condenser water supply temperatures are denoted by 1-2-3-4 and 7-2-5-6, respectively. In Figure 2.19, it is evident that the decrease in condenser water supply temperature leads to a reduction in the difference between the condenser and evaporator pressures, shifting from $(P_{cond,3} - P_{eva})$ to $(P_{cond,5} - P_{eva})$. Additionally, the cooling effect per unit mass of flowing refrigerant increases from $(h_2 - h_1)$ to $(h_2 - h_7)$ due to this temperature decrease. Consequently, the refrigerant flow rate through the refrigeration cycle decreases to maintain the same cooling effect. This results in reduced compressor power consumption and an overall improvement in the chiller's energy efficiency. Similar to the chilled water supply temperature, typically,

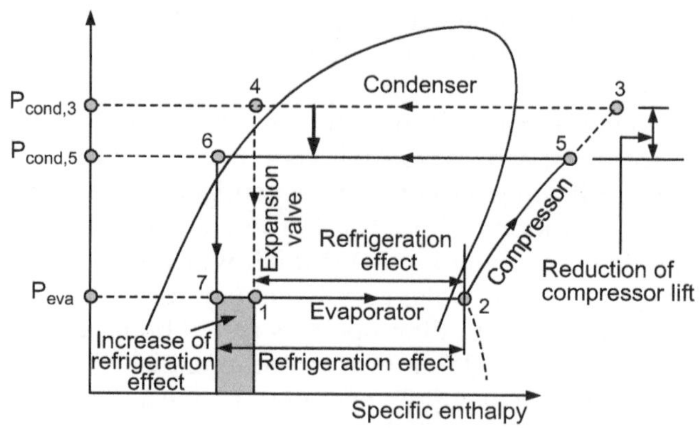

FIGURE 2.19
Effect of condenser water temperature reset.

a decrease in the condenser water supply temperature by 1°C can result in an enhancement of chiller efficiency by approximately 3% [5]. This improvement is attributed to the reduction in compressor lift and the increase in the cooling effect. However, it is worth noting that the decrease in condenser water supply temperature leads to increased power consumption by the cooling tower. To determine the net energy saving, careful analysis of the increase in cooling tower power consumption versus the reduction in chiller compressor power consumption is necessary.

2.6 Case Study: Design of Condenser Water Pumping System

The cooling water is circulated from the basin of a cooling tower through three centrifugal condenser water pumps connected in parallel to three chillers, as depicted in Figure 2.20. The pumps, motors, and couplings have efficiencies of 70%, 90%, and 98%, respectively. Each chiller outlet is equipped with a motorized modulating valve, which regulates the flow of cooling water by adjusting the valve opening via its motor.

Various components such as valves, strainers, and fittings are installed in the piping system. To simplify and clearly present the condenser water loop's piping layout, these details are omitted from Figure 2.20. Instead, different sections of the piping system are labeled with names like section-AB and section-CD. Table 2.9 provides the required cooling water flow rates and the corresponding pressure losses for the chiller condensers. Table 2.10 summarizes the details of the various pipe sections and their associated fittings. The flow rate of cooling water in different sections of the piping system is calculated by summing the flow through the chillers and the respective branches of the piping system.

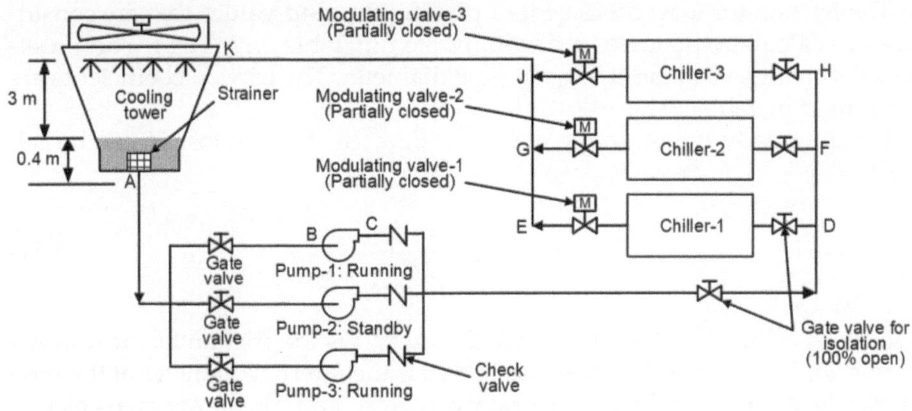

FIGURE 2.20
Piping layout of condenser water loop of a chiller system.

TABLE 2.9
Cooling Water Flow Rates and Pressure Losses in the Condensers of Chillers

Heat Exchanger of Condenser	Cooling Water Flow Rate, (m^3/s)	Pressure Loss (m)
Chiller 1	0.075	10
Chiller 2	0.08	8
Chiller 3	0.085	6

TABLE 2.10
The Details of the Different Pipe Sections and Their Associated Fittings

Pipe Section	Flow, m^3/s	Pipe Dia., m	Pipe Length, m	No. of 100% Open Gate Valves	No. of Check Valves	No. of Strainers	No. of Modulating Valves	No. of 90° Long Radius Bends	No. of Tees
AB	0.240	0.3	15	2	0	1	0	8	0
CD	0.240	0.3	35	2	1	0	0	8	0
DE	0.075	0.2	5	4	0	0	1	6	2
FG	0.080	0.2	5	4	0	0	1	6	2
HJ	0.085	0.2	5	4	0	0	1	5	0
DF	0.165	0.25	35	0	0	0	0	3	1
FH	0.085	0.2	25	0	0	0	0	3	1
JG	0.085	0.2	25	0	0	0	0	3	0
GE	0.165	0.25	35	0	0	0	0	3	1
JK	0.240	0.3	45	2	0	0	0	7	1

The friction (or loss) coefficient of pipe fittings and values that are considered to calculate the losses are summarized in Table 2.11. The friction coefficients change with the change in pipe diameter. The friction coefficients are presented in Table 2.11.

For the steady flow through a circular pipe, the friction loss H_f can be calculated as

$$H_f = \frac{4fL}{D}\left(\frac{v^2}{2g}\right) \tag{2.21}$$

where H_f is the head loss due to friction in m, f is the friction factor (dimensionless), L is the length of the pipe in m, d is the internal diameter of the pipe in m, g is the acceleration due to gravity in m/s^2, and v is the average velocity of flow in m/s. The average velocity of flow can be calculated as

$$v = \frac{\text{Average volume flow rate}\left(\frac{m^3}{s}\right)}{\text{Cross}-\text{sectional area of pipe}\left(m^2\right)} \tag{2.22}$$

Head loss across a valve/fitting can be expressed as

$$h_f = \alpha\left(\frac{v^2}{2g}\right) \tag{2.23}$$

where h_f is the head loss due to valve/fitting in m and α is the friction, loss, or resistance coefficient. The input power to the motor of the pump can be expressed as

$$P_{\text{input}} = \frac{Q \times H_{\text{total}} \times \rho \times g}{1000 \times \eta_{\text{motor}} \times \eta_{\text{coupling}} \times \eta_{\text{pump}}} \tag{2.24}$$

where P_{input} is the input power to the motor of the pump in kW, H_{total} is the total head loss of the piping system in m, ρ is the density of water in kg/m^3,

TABLE 2.11

Friction (or Loss) Coefficient of Pipe Fittings and Values [6]

Pipe Diameter, mm	90° Ellow Long	Gate Valve (Fully Open)	Check Valve	Tee-Line	Tee-Branch	Strainer
200	0.16	0.08	2.0	0.10	0.58	2.2
250	0.14	0.06	2.0	0.09	0.53	2.2
300	0.13	0.05	2.0	0.08	0.50	2.2

η_{motor} is the efficiency of the motor, $\eta_{coupling}$ is the efficiency of the coupling, and η_{pump} is the efficiency of the pump.

Sample Calculation for Piping Section-AB

Length of section-AB, $L = 15$ m

Diameter of section-AB, $D = 300$ mm $= 0.3$ m

Volume flow rate of water, $Q = 240$ L/s $= 0.24$ m³/s

Water flow velocity, $v = Q/A = \dfrac{Q}{\frac{1}{4}\times\pi\times D^2} = \dfrac{0.24}{\frac{1}{4}\times\pi\times 0.3^2} = 3.397\,\text{m/s}$

Assume friction factor for the pipe, $f = 0.006$

Friction loss in pipe $= H_f = \dfrac{4fL}{D}\left(\dfrac{v^2}{2g}\right) = \dfrac{4\times 0.006\times 15}{0.3}\times\dfrac{3.397^2}{2\times 9.81} = 0.706\,\text{m}$

Velocity head, $H_{vel} = \dfrac{v^2}{2g} = \dfrac{3.397^2}{2\times 9.81} = 0.588\,\text{m}$

No. of 100% open gate valves = 2

The loss coefficient for the gate valve = 0.05

Head loss for gate valve, $h_{\text{gate valve}} = \alpha\left(\dfrac{v^2}{2g}\right) = 2\times 0.05\times 0.588 = 0.059\,\text{m}$

No. of strainers = 1

Loss coefficient for strainers = 2.2

Head loss for strainers, $h_{\text{strainer}} = \alpha\left(\dfrac{v^2}{2g}\right) = 1\times 2.2\times 0.588 = 1.294\,\text{m}$

No. of 90° bends = 8

Loss coefficient for 90° bends = 0.13

Head loss for 90° bends $= h_{90^\circ\text{bends}} = \alpha\left(\dfrac{v^2}{2g}\right) = 8\times 0.13\times 0.588 = 0.612\,\text{m}$

Total head losses across section-AB $= H_f + h_{\text{gate valve}} + h_{\text{strainer}} + h_{90^\circ\text{bends}} = 2.67\,\text{m}$

Details of different sections of the piping system and their corresponding head losses are summarized in Table 2.12.

Total head loss in the piping loop A-B-C-D-E-K

= 2.67 + 3.49 + 16.12 + 2.76 = 25 m

Total head loss in the piping loop A-B-C-D-F-G-E-K

= 2.67 + 3.49 + 11.65 + 2.23+ 2.23 + 2.76 = 25 m

Total head loss in the piping loop A-B-C-D-F-H-J-G-E-K

= 2.67 + 3.49 + 9.03 + 2.23+ 1.34 + 1.30 + 2.23 + 2.76 = 25 m

TABLE 2.12

Different Sections of the Piping System and Their Corresponding Head Losses

Pipe Section	Flow, m^3/s	Pipe Diameter, m	Water Flow Velocity, m/s	Pipe Length, m	Friction Factor	Friction Loss in Pipe, m	$V^2/(2g)$	No. 100% Open Gate Valves	Loss Coefficient for Gate Valve	Loss for Gate Valves, m
AB	0.240	0.3	3.397	15	0.006	0.706	0.588	2	0.05	0.059
CD	0.240	0.3	3.397	35	0.006	1.647	0.588	2	0.05	0.059
DE	0.075	0.2	2.389	5	0.006	0.174	0.291	4	0.08	0.093
FG	0.080	0.2	2.548	5	0.006	0.199	0.331	4	0.08	0.106
HJ	0.085	0.2	2.707	5	0.006	0.224	0.373	4	0.08	0.120
DF	0.165	0.25	3.363	35	0.006	1.937	0.576	0	0	0.000
FH	0.085	0.2	2.707	25	0.006	1.120	0.373	0	0	0.000
JG	0.085	0.2	2.707	25	0.006	1.120	0.373	0	0	0.000
GE	0.165	0.25	3.363	35	0.006	1.937	0.576	0	0	0.000
JK	0.240	0.3	3.397	45	0.006	2.117	0.588	2	0.05	0.059

TABLE 2.12 (CONTINUED)

Different Sections of the Piping System and Their Corresponding Head Losses

Pipe Section	No. of Check Valves	Loss Coefficient for Check Valve	Loss for Check Valves (m)	No. Modulating Valves	Loss Coefficient of Partially Closed Modulating Valves	Loss for Modulating Valves (m)	No. of Strainers	Loss Coefficient for Strainer	Loss for Strainer (m)	No. 90° Bends	Loss Coefficient for 90° Bend	Bend Loss, m
AB	0	0	0.000	0	0	0	1	2.2	1.294	8	0.13	0.612
CD	1	2.000	1.176	0	0	0	0	0	0.000	8	0.13	0.612
DE	0	0	0.000	1	18	5.234	0	0	0.000	6	0.16	0.279
FG	0	0	0.000	1	8	2.647	0	0	0.000	6	0.16	0.318
HJ	0	0	0.000	1	5.8	2.166	0	0	0.000	5	0.16	0.299
DF	0	0	0.000	0	0	0	0	0	0.000	3	0.14	0.242
FH	0	0	0.000	0	0	0	0	0	0.000	3	0.16	0.179
JG	0	0	0.000	0	0	0	0	0	0.000	3	0.16	0.179
GE	0	0	0.000	0	0	0	0	0	0.000	3	0.14	0.242
JK	0	0	0.000	0	0	0	0	0	0.000	7	0.13	0.535

(Continued)

TABLE 2.12 (CONTINUED)

Different Sections of the Piping System and Their Corresponding Head Losses

Pipe Section	No. Tees	Tee Loss Coefficient	Tee Loss (m)	Chillers (m)	Total Loss (m)	Pipe
AB	0	0	0.000	0	2.67	AB
CD	0	0	0.000	0	3.49	CD
DE	2	0.58	0.337	10	16.12	DE
FG	2	0.58	0.384	8	11.65	FG
HJ	1	0.58	0.217	6	9.03	HJ
DF	1	0.09	0.052	0	2.23	DF
FH	1	0.1	0.037	0	1.34	FH
JG	0	0	0.000	0	1.30	JG
GE	1	0.09	0.052	0	2.23	GE
JK	1	0.08	0.047	0	2.76	EK

Therefore, the total friction head for each piping loop, $H_{friction} = 25$ m

Static head for the cooling tower, $H_{static} = 3$ m

Total head developed by the pump, $H_{total} = H_{friction} + H_{static} = 25 + 3 = 28$ m

Total water flow rate = Water flow rate through chiller 1 + chiller 2 + chiller 3 = 0.075 + 0.08 + 0.085 = 0.24 m^3/s

The pumping system consists of three pumps. Two pumps are running and one pump is on standby.

Therefore, the water flow rate through each pump, $Q = (0.24/2) = 0.12$ m^3/s

The power consumption of each pump:

$$P_{each} = \frac{Q \times H_{total} \times \rho \times g}{1000 \times \eta_{motor} \times \eta_{coupling} \times \eta_{pump}}$$

$$P_{each} = \frac{0.12 \times 28 \times 1000 \times 9.81}{1000 \times 0.9 \times 0.98 \times 0.7} = 53.4\,\text{kW}$$

Total power consumption for the two operating pumps:

$$P_{total} = 53.4 \times 2 = 106.8\,\text{kW}$$

2.7 Chilled Water and Condenser Water Pumping System

Chilled water and condenser water pumping systems are used in water-cooled central air-conditioning chilled water systems to circulate the

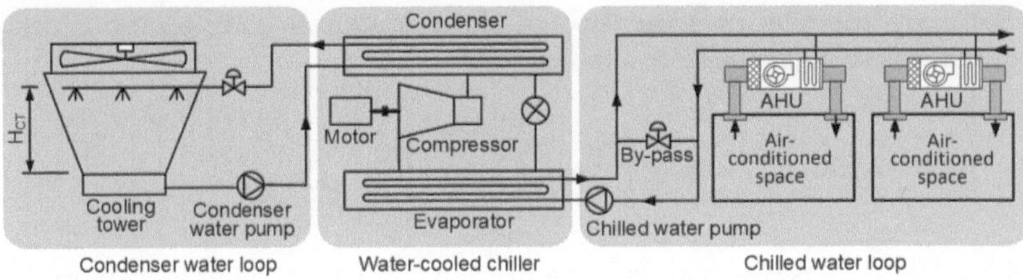

FIGURE 2.21
Chilled water loop and condenser water loop of water-cooled chiller system.

chilled water through AHUs and the condenser water through the cooling towers, respectively, as depicted in Figure 2.21. The chilled water pumping system is a closed-loop system as the chilled water is enclosed within the piping networks and is not exposed to the atmosphere. Chilled water pumps need to overcome the frictional losses caused by different components within the closed loops such as pipe sections, heat exchangers of the evaporators of chillers, cooling coils of AHUs, bends, valves, strainers, sensors, and other fittings. The condenser water pumping system, on the other hand, is an open loop system as the condenser water is sprayed in the open cooling towers. Condenser water pumps need to develop an additional static pressure head equivalent to the vertical height of the open sections to cooling towers together with the frictional losses caused by different components within the open loops such as pipe sections, heat exchangers of the condensers of chillers, bends, valves, strainers, sensors, and other fittings.

The total pressure drop of a closed loop-n of a chilled water pumping system can be calculated as

$$\Delta P_{\text{chw,loop-n}} = \left(\frac{8\rho}{\pi^2}\right)\left[\sum_{i=1}^{n1}\left(\frac{fL}{D^5}Q^2\right)_i + \sum_{j=1}^{n2}\left(\frac{C_o}{D^4}Q^2\right)_j\right] \tag{2.25}$$

where $\Delta P_{\text{chw,loop-n}}$ is the total pressure drop in chilled water loop-n in Pa, f is the Moody dimensionless friction factor, L is the length of the pipe section in m, D is the inner diameter of the pipe in m, V is the average velocity of the fluid in m/s, ρ is the density of chilled water in kg/m^3, C_o is the dimensionless loss coefficient of the fitting such as elbow, tee-joint, valve, and strainer, Q is the volume flow rate of water in m^3/s, n1 is the number of pipe sections in loop-n, and n2 is the number of fittings in loop-n. The first term of Eq. (2.25) represents the friction losses of straight pipe sections, and the second term represents the pressure losses for the fittings.

Similarly, the total pressure drop of an open loop-m of condenser water pumping system can be calculated as

$$\Delta P_{\mathrm{cw,loop-m}} = H_{\mathrm{CT}}\rho g + \left(\frac{8\rho}{\pi^2}\right)\left[\sum_{i=1}^{m1}\left(\frac{fL}{D^5}Q^2\right)_i + \sum_{j=1}^{m2}\left(\frac{C_o}{D^4}Q^2\right)_j\right] \tag{2.26}$$

where $\Delta P_{\mathrm{cw,loop\text{-}m}}$ is the total pressure drop in condenser water loop-m in Pa, H_{CT} is the vertical height of cooling tower in m, ρ is the density of condenser water in kg/m^3, g is the acceleration due to gravity in m/s^2, m1 is the number of pipe sections in loop-m, and m^2 is the number of fittings in loop-m. The first term of Eq. (2.26) represents the static pressure head corresponding to the open height of the cooling tower, the second term denotes the friction losses of straight pipe sections, and the third term represents the pressure losses for the fittings.

The function of a pump is to generate the necessary pressure head to overcome the total pressure loss of the water loop of maximum pressure drop and circulate the required amount of water through all piping loops. The input electric power to the motor of a chilled water pump can be calculated as

$$W_{\mathrm{chwp,input}} = \frac{\Delta P_{\mathrm{chw,loop},max}\, Q_{\mathrm{chw,total}}}{\eta_{\mathrm{motor}}\eta_{\mathrm{coupling}}\eta_{\mathrm{pump}}} \tag{2.27}$$

where $W_{\mathrm{chwp,input}}$ is the input electric power to the motor of the chilled water pump in W, $\Delta P_{\mathrm{chw,loop,max}}$ is the pressure loss of the chilled water piping loop of maximum pressure drop in Pa, $Q_{\mathrm{chw,total}}$ is the total volume flow rate of chilled water through all chilled water loops in m^3/s, η_{motor} is the efficiency of the motor, η_{coupling} is the efficiency of the coupling system, and η_{pump} is the efficiency of the pump.

Similarly, the input electric power to the motor of a condenser water pump can be calculated as

$$W_{\mathrm{cwp,input}} = \frac{\Delta P_{\mathrm{cw,loop},max}\, Q_{\mathrm{cw,total}}}{\eta_{\mathrm{motor}}\eta_{\mathrm{coupling}}\eta_{\mathrm{pump}}} \tag{2.28}$$

where $W_{\mathrm{cwp,input}}$ is the input electric power to the motor of the condenser water pump in W, $\Delta P_{\mathrm{cw,loop,max}}$ is the pressure loss of the condenser water piping loop of maximum pressure drop in Pa, $Q_{\mathrm{cw,total}}$ is the total volume flow rate of condenser water through all condenser water loops in m^3/s, η_{motor} is the efficiency of the motor, η_{coupling} is the efficiency of the coupling system, and η_{pump} is the efficiency of the pump.

Eqs. (2.25) and (2.26) show that the pressure drop of a piping loop is proportional to the square of the water flow rate through the piping loop. Therefore, the plot of the pressure drops of a piping system versus the volume flow rate of water, known as the system curve, is approximately parabolic. The system

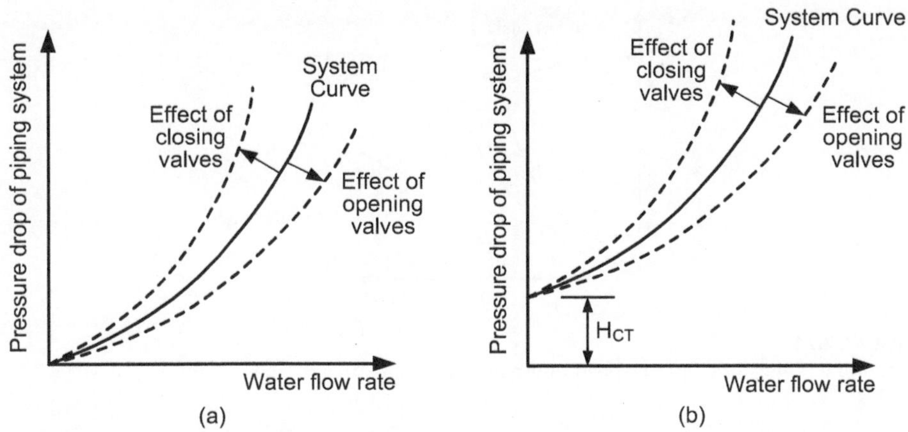

FIGURE 2.22
System curves for the chilled water loop and the condenser water loop. (a) System curve for chilled water loop and (b) system curve for condenser water loop.

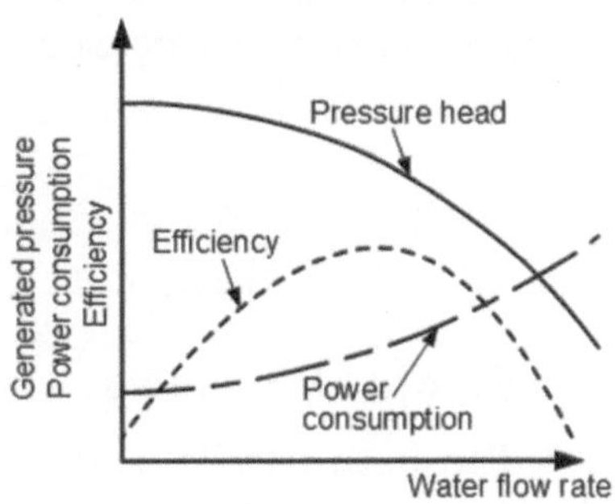

FIGURE 2.23
Pump characteristic curve.

curves for the chilled water loop and the condenser water loop are shown in Figure 2.22.

The variations in pressure head generation capacity, electric power consumption, and efficiency of a pump with changes in volume flow rates of water are known as the pump characteristic curve, illustrated in Figure 2.23. The pressure head generation curve shifts downward when the impeller speed of a pump decreases. The system characteristic curve of a piping system is plotted on the pump characteristic curve. The intersection of the system characteristic curve and the pump characteristic curve represents the operating point as shown in Figure 2.24 [7]. An appropriately selected pump operates within the best efficiency region most of the time. During the period of low cooling load in air-conditioned spaces, the flow rate of chilled water and condenser water delivered by

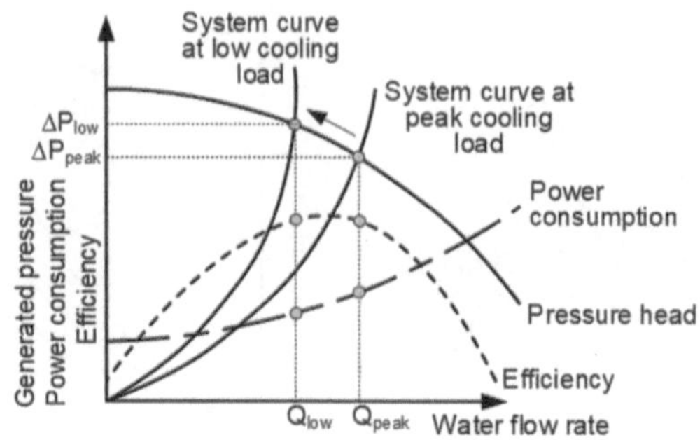

FIGURE 2.24
Operating points for constant speed pump at different loadings.

the constant speed pumps can be reduced by partially closing the flow-modulating valves. The operating points for peak and low cooling loads are illustrated in Figure 2.24. When the flow rate of water is reduced by partially closing the flow-modulating valves, the pressure head generated by the constant speed pumps increases. Consequently, the electric power consumption of the pumps remains nearly constant during the period of low cooling loads.

2.8 Parallel and Series Pumping Systems

Pumps are often connected in both parallel and series configurations to accommodate variable flow of water, overcome total pressure losses in the piping network, and ensure redundancy in case of any pump failure [8]. As illustrated in Figure 2.7, three primary chilled water pumps and three secondary chilled water pumps are connected in a parallel configuration. However, the secondary chilled water pumps are arranged in series with the primary chilled water pumps. Parallel configurations are used to achieve a higher flow rate of water. In such setups, each pump generates an equal pressure head, and the total flow rate is the sum of the flow rates of all operating pumps. Conversely, in series configurations, the total flow rate of water through each set of pumps connected in series remains the same. However, the pressure head increases after each set of pumps. Thus, parallel pumping systems are designed to meet high flow rate requirements, while series configurations are ideal for piping systems with high-pressure losses. The operating points, flow rates, and pressure heads generated by each pump in parallel and series configurations are illustrated in Figure 2.25.

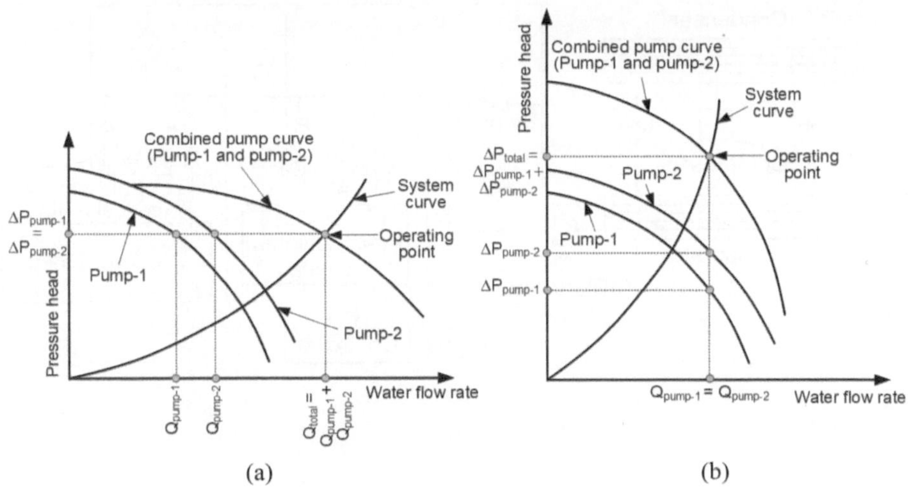

FIGURE 2.25
Operating points, flow rates, and pressure heads generated by each pump for the parallel and series pumping configurations. (a) Parallel pumping system and (b) series pumping system.

2.9 Control Strategies for Energy-Efficient Pumping Systems

During the period of low cooling loads, the flow rate of chilled water and condenser water can be reduced using variable-speed pumps with appropriate control strategies as illustrated in Figures 2.26 and 2.27. When the cooling load in air-conditioned spaces is low, the flow-modulating valves located at the outlet of each AHU are partially closed to restrict the flow of chilled water and reduce the cooling effect. This causes an increase in water pressure in the piping network, which is measured by a differential pressure sensor that provides feedback to the controller as illustrated in Figure 2.26(a). The controller then reduces the pump speed using a VSD to maintain a pre-set constant differential pressure, resulting in significant energy savings as depicted in Figure 2.26(b). If the differential pressure sensor is installed in the plant room near the pump, the pressure difference across the pump remains constant during both full and part-load operations. However, if the sensor is located across the furthest critical AHU, the pressure difference across the pump decreases under part-load operation due to the reduction of piping losses at low flow rates, leading to further energy savings. The flow-modulating valves regulate the flow of chilled water in each AHU, ensuring that comfort conditions in each air-conditioned space are accurately maintained. However, the capital cost of the AHU system increases due to the installation of flow-modulating valves in each AHU.

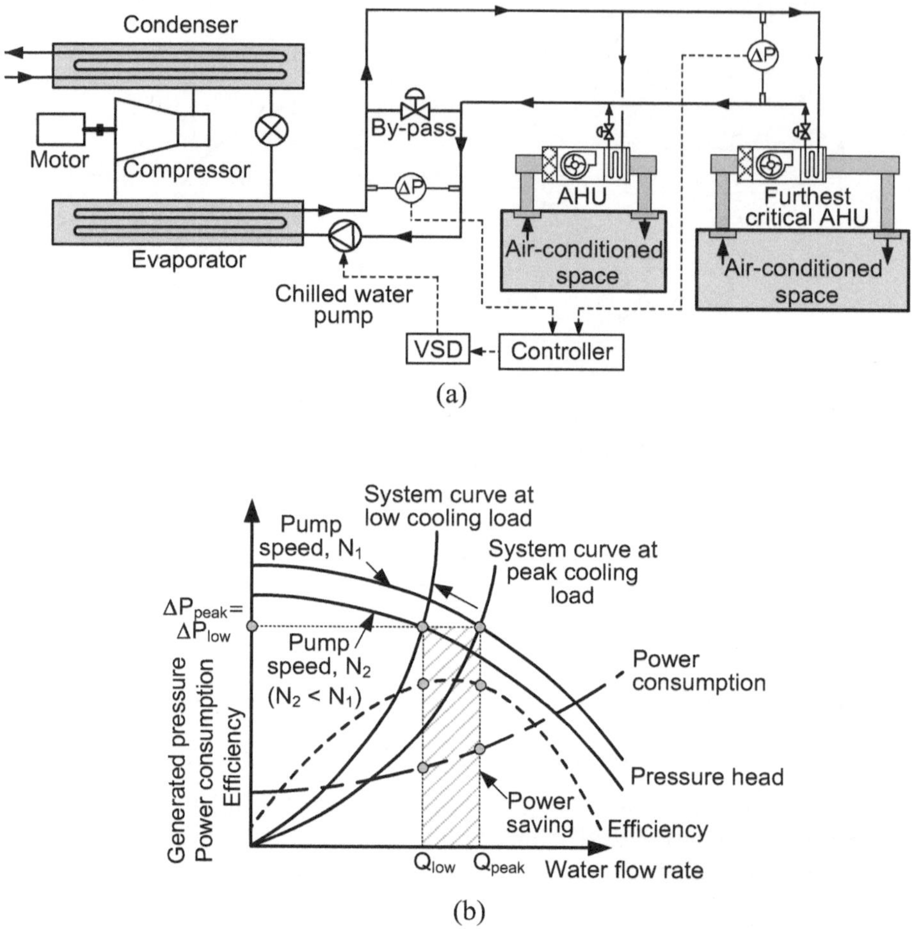

FIGURE 2.26
Control of chilled water pump based on differential pressure: (a) feedback control system on feedback of differential pressure and (b) reduction of pump power consumption.

For large open spaces such as a retail mall, the cooling load of each AHU may follow the same cooling load profile. In such applications, the speed of the chilled water pumps can be modulated based on the chilled water return temperature as shown in Figure 2.27(a). When the cooling load decreases, the chilled water return temperature drops below the pre-set value. The controller then reduces the pump speed to decrease the flow rate of chilled water aligning with the reduced cooling load. Since this system does not use flow-modulating valves, the system curve remains unchanged. The pump's operating point moves along the system curve with changes in pump speed, as illustrated in Figure 2.27(b), resulting in a reduction in both water flow rate and system pressure. This approach allows for maximum energy savings

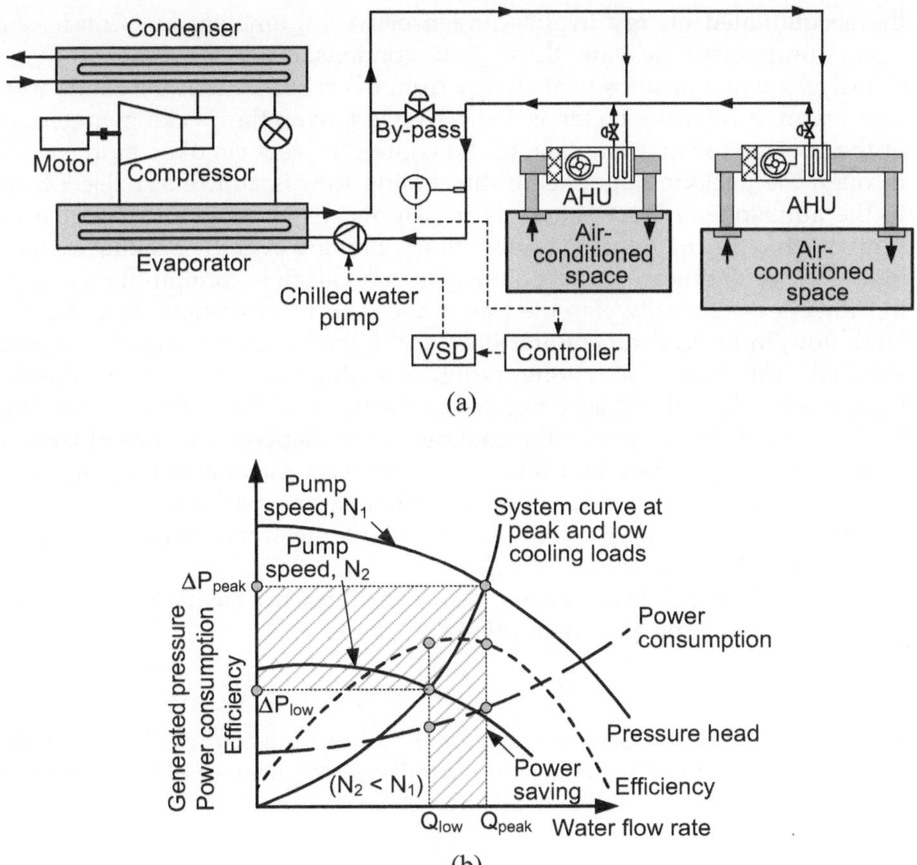

FIGURE 2.27
Control of chilled water pump based on chilled water return temperature: (a) control system based on feedback of chilled water return temperature and (b) reduction of pump power consumption.

during part-load operation. However, if the cooling load of the spaces does not follow the same profile, maintaining comfort conditions in each space could be challenging.

2.10 Cooling Tower System

The heat energy absorbed by the evaporators from the air-conditioned spaces and the electrical energy supplied to the compressors to carry out the compression of refrigerant vapor are finally transferred to the condensers of the water-cooled chillers. The function of the cooling towers is to reject the accumulated energy in the condenser to the ambient air. Condenser water pumps circulate water through the condensers and the cooling towers. Condenser water absorbs heat energy from the refrigerant in the condenser. The warm condenser water is then sprayed over the packing materials of the cooling towers. The fans of the cooling towers circulate ambient air through the packing materials of the cooling towers and finally reject heat to the atmosphere. Heat energy is transferred as sensible and latent heat from the thin film of warm condenser water flowing over the packing materials to the circulating air. Based on the water and air flow configurations, cooling towers are broadly classified as induced-draft cross-flow, forced-draft cross-flow, induced-draft counter-flow, and forced-draft counter-flow types. Induced-draft cross-flow configuration is widely used in cooling towers. Finally, the cold water is accumulated in the basin of the cooling tower. The heat transfer performance of the cooling towers depends on several factors such as the temperature and relative humidity of the ambient air, air flow rate, water flow rate, condition of water spray, and packing materials [9]. Air and water flow configuration and the heat transfer mechanisms are illustrated in Figure 2.28.

The rate of sensible heat transfer from a small control element of the water film to the circulating air is determined as

$$Q_s = dA\left(T_w - T_a\right)h_c \tag{2.29}$$

where Q_s is the rate of sensible heat transfer from the control element in W, dA is the exposed surface area of the small control element of water film in

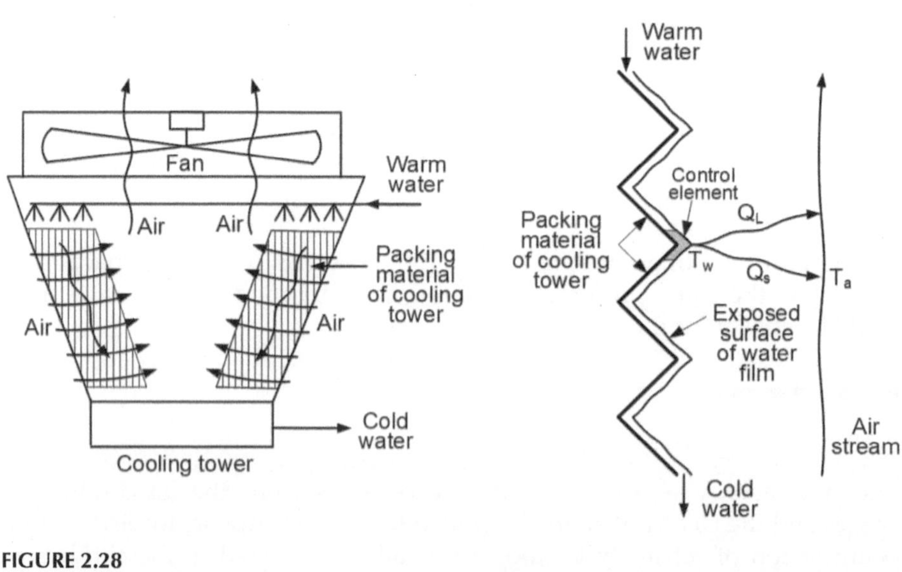

FIGURE 2.28
Air and water flow configuration and the heat transfer mechanisms of an induced-draft cross-flow cooling tower.

m², h_c is the convective heat transfer coefficient of air in W/m² K, T_w = temperature of water film of the control element in °C, and T_a is the temperature of circulating air in °C.

A small fraction of the flowing water evaporates by absorbing the latent heat of vaporization from the water and the sensible heat from the circulating air. Finally, the evaporated water is exhausted to the ambient. The temperature of the water film drops as the evaporation process takes place by absorbing the latent heat of vaporization from the water film. The latent heat transfer rate from a small control element of the water film can be calculated as

$$Q_L = \dot{m}_v dA h_{fg} \tag{2.30}$$

where Q_L is the rate of latent heat transfer from the control element in W, $\dot{m}_v$ is the mass flux of evaporated water in kg/m² s, and h_{fg} is the latent heat of vaporization at T_w in J/kg. The mass flux of evaporated water is evaluated as

$$\dot{m}_v = \left(m_{v,f} - m_{v,a}\right) k_m \tag{2.31}$$

where k_m is the mass transfer coefficient in kg/m² s, $m_{v,f}$ is the mass fraction of water vapor at the exposed surface of the water film (dimensionless), and $m_{v,a}$ is the mass fraction of water vapor in the circulating air (dimensionless). The convective heat transfer coefficient h_c and the mass transfer coefficient depend on several factors such as airflow velocity, flow geometry, temperature, and thermophysical properties of air. The exposed surface of the water film is considered as saturated. The mass fraction of water vapor at the exposed surface of the water film is determined as

$$m_{v,f} = \frac{\left(P_{\text{sat at } T_w}\right) M_v}{\left(P_{\text{sat at } T_w}\right) M_v + P_{a,f} M_a} \tag{2.32}$$

where $P_{\text{sat at } Tw}$ is the saturated vapor pressure at T_w in Pa, $P_{a,f}$ is the partial pressure of air at the exposed surface of water film in Pa, M_a is the molecular weight of air in kg/kmol, and M_v is the molecular weight of vapor in kg/kmol. Similarly, the mass fraction of water vapor in the air stream can be calculated as

$$m_{v,a} = \frac{(\text{RH})\left(P_{\text{sat at } T_a}\right) M_v}{(\text{RH})\left(P_{\text{sat at } T_a}\right) M_v + P_a M_a} \tag{2.33}$$

where $P_{\text{sat at } Ta}$ is the saturated vapor pressure at T_a in Pa, RH is the relative humidity of air in fraction, and P_a is the partial pressure of air in the air stream in Pa. It is evident from Eq. (2.29) that the sensible heat transfer rate depends on the temperature of the surrounding ambient air.

As the ambient air temperature increases, the sensible heat transfer rate decreases, eventually reaching zero if the air temperature equals the water film temperature. Eqs. (2.30)–(2.33) indicate that the evaporation rate of water, and consequently the latent heat transfer rate from the water film, depends on the relative humidity of the ambient air. The latent heat transfer rate decreases as the relative humidity increases. When the relative humidity reaches 100% (saturation), both the water evaporation rate and the corresponding latent heat transfer rate drop to zero. The total heat rejection rate of a cooling tower is the sum of the sensible and latent heat transfer rates.

In an ideal scenario, the temperature of water exiting a cooling tower could be reduced to the wet bulb temperature of the air. However, achieving this would require an impractically large cooling tower, making it economically unfeasible. Typically, cooling towers are designed with an approach temperature of about 2.8°C. The approach temperature is the difference between the temperature of the water leaving the cooling tower and the wet-bulb temperature of the surrounding air.

Cooling towers reject the heat of the air-conditioning spaces along with the heat added by the compressor to the refrigerant vapor during the compression process. The amount of heat added by the compressor depends on its efficiency. Typically, the heat added by the compressor system is about 25% of the chiller's cooling capacity. Thus, a cooling tower's capacity should be approximately 125% of the chiller's cooling capacity. A general rule is to select a cooling tower with a capacity of 1.5 times that of the chiller. An oversized cooling tower, with its greater heat transfer surface area, can reject the same amount of heat with less air circulation, reducing the power consumption of the cooling tower fans. Additionally, it can lower the water temperature more effectively, improving chiller efficiency. However, the initial cost of an oversized cooling tower is higher. Therefore, sizing a cooling tower involves balancing the higher upfront costs against the potential operating cost savings for both the cooling towers and the chillers.

The cooling load of air-conditioned spaces fluctuates over time, as illustrated in Figure 2.29. Cooling towers are typically selected based on the cooling capacity of the chillers. To optimize the power consumption of cooling towers, their capacity can be adjusted according to the actual dynamic cooling load of the chillers. This modulation can be achieved by varying the air flow rate through fan staging or by using VSDs to regulate fan speed. In fan staging systems, several cooling tower fans are turned on or off based on the dynamic heat rejection load to maintain the pre-set temperature of the condenser water leaving the cooling towers. In variable-speed fan systems, the fan speed is adjusted with VSDs to achieve the same temperature control. Additionally, fan speed can be modulated based on a pre-set approach temperature, which is determined by measuring the difference between the ambient air's wet-bulb temperature and the temperature of the condenser water leaving the cooling towers, as shown in Figure 2.29. The affinity laws

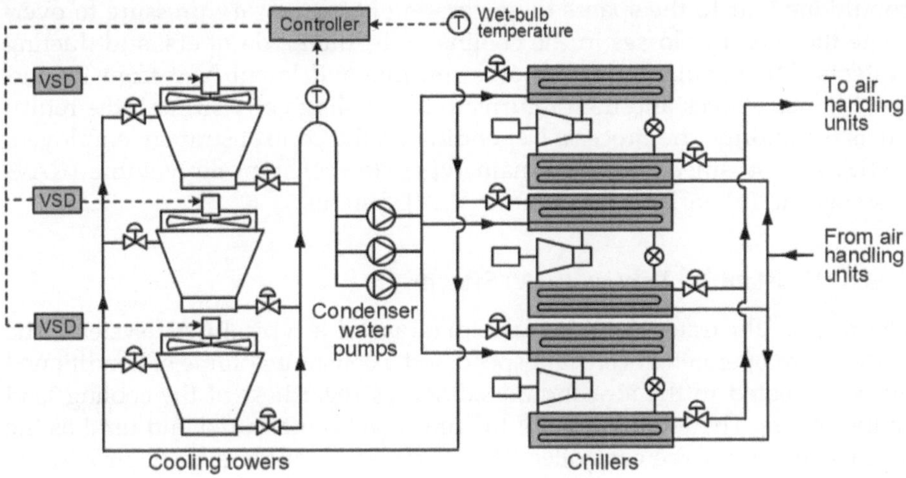

FIGURE 2.29
Control of cooling tower capacity based on approach temperature.

are often used to estimate power consumption at lower fan speeds. According to these laws, the power consumption of a fan is proportional to the cube of its speed. Therefore, if the heat rejection load of a cooling tower decreases to 80%, the fan's power consumption will drop to about 50% ($0.8^3 = 0.512$). This means that variable-speed fan systems can significantly reduce cooling tower fan power consumption under part-load conditions. Typically, chiller plants are equipped with standby cooling towers. By operating all cooling towers, including the standby units, at lower speeds, the overall power consumption of the fans can be further reduced.

2.11 Air Handling Systems

The comfort conditions of air-conditioned spaces are maintained by cooling and dehumidifying the circulating air using AHUs. The main components of an AHU include blowers, cooling coils, filters, dampers, and a ducting system. Chilled water pumps circulate chilled water from the chiller evaporators to the cooling coils of AHUs installed at strategic locations throughout the building. A mixture of outdoor fresh air and return air is circulated over the cooling coil by the blower. This process cools and dehumidifies the air by transferring sensible and latent heat to the chilled water in the cooling coils. The warmed water is then returned to the chiller evaporators for cooling, continuing the chilled water circulation cycle. Blowers distribute the conditioned air to the spaces by generating the necessary pressure to overcome the pressure losses in the cooling coils, filters, dampers, and ducting systems. The circulating air absorbs sensible and latent heat from the air-conditioned spaces and then returns to the cooling coils through the return ducts to continue the process. Depending on the control strategy employed, AHUs are classified into two main types: (a) constant air volume (CAV) systems and (b) variable air volume (VAV) systems.

2.11.1 Constant Air Volume (CAV) Systems

Figure 2.30 illustrates the schematic diagram of a typical CAV system. The blowers are operated at constant speed and a constant volume of conditioned air is circulated to the air-conditioned spaces regardless of the cooling load of the spaces. The temperature of the return air is measured and used as the input variable for the controller. The temperature of the supply air and the flow rate of the chilled water through the coil of the AHU coil are modulated by the controller in response to the measured return air temperature which varies with the change of cooling load of the air-conditioned spaces. A CO_2 sensor installed in the return duct measures the overall CO_2 level in the spaces. This measured CO_2 level serves as the input variable for the controller, which adjusts the openings of the return and fresh air motorized dampers to ensure that only the minimum required amount of outdoor fresh air is drawn from the atmosphere and maintains the recommended CO_2 level in spaces. Since atmospheric air is generally warm and humid, drawing in excess fresh air would increase the cooling load on the AHUs and chillers. CAV systems are ideal for large and open areas where cooling load variations are relatively uniform.

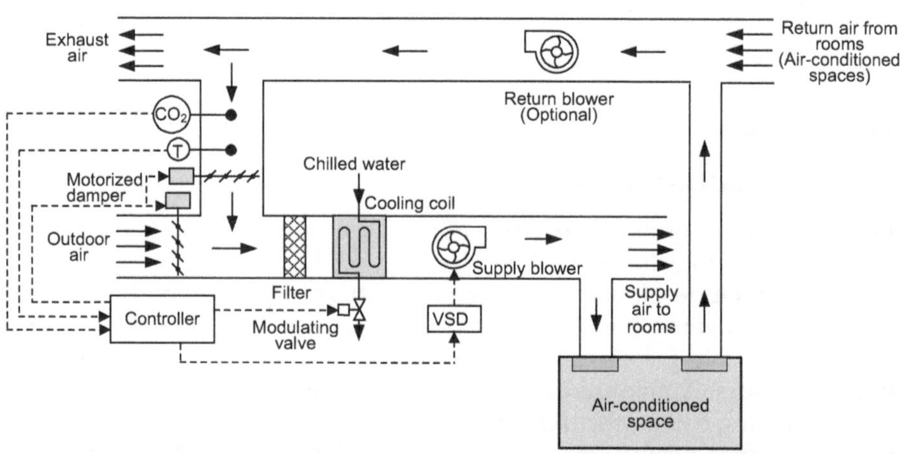

FIGURE 2.30
Schematic diagram of a typical constant air volume AHU system.

However, during low part-load conditions, when less chilled water is supplied to the AHU coil, dehumidification of the supply air may be reduced, potentially increasing the relative humidity in the conditioned spaces.

2.11.2 Variable Air Volume (VAV Systems

In variable air volume (VAV) systems, the airflow rate to air-conditioned spaces is adjusted based on the cooling load. Figure 2.31 illustrates the schematic diagram of a typical VAV system. The temperature of the supply air is kept constant regardless of the cooling load, while the chilled water flow rate through the AHU coil is modulated according to the cooling demand. This coordinated modulation of the circulating air and the flow rate of chilled water ensures the necessary dehumidification of the supply air at low part-load conditions. VAV systems can effectively maintain required comfort conditions in different spaces even if the cooling load varies unevenly in each space. The blower can be operated at either a constant or variable speed. Airflow in air-conditioned spaces can be adjusted in response to changes in cooling load using different methods such as discharge dampers, inlet guide vanes, and VSDs. Consequently, the power consumption of the blower varies with the cooling load. Under part-load conditions, the reduction in blower power consumption using discharge dampers and inlet guide vanes is significantly less compared to VSD systems, which closely follow the "cubic" blower power relationship described by the affinity law.

Each air-conditioned space is equipped with a motorized damper, known as a VAV box, which modulates airflow. The temperature sensor together with the controller in each space controls the damper opening and the volume flow rate of

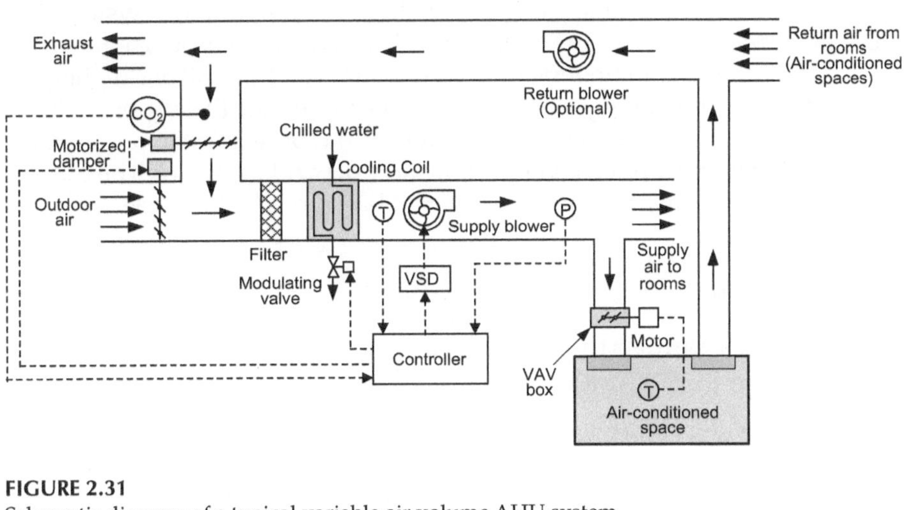

FIGURE 2.31
Schematic diagram of a typical variable air volume AHU system.

air based on a pre-set temperature. When the cooling load is low and the motorized damper is partially closed, the air pressure in the main duct increases. This increase is detected by a static pressure sensor in the main duct. Additionally, the discharge air temperature is monitored by a temperature sensor in the main duct. The controller uses feedback from both the static pressure sensor and the temperature sensor to regulate airflow. It adjusts the blower speed using a VSD and modulates the chilled water flow through the cooling coil by controlling a modulating valve. During part-load conditions, the blower speed decreases in proportion to the reduced airflow, resulting in a significant drop in blower power consumption. The chilled water supply temperature can also be raised during the period of low cooling load which will decrease the compressor lift and reduce the electric power consumption of the chiller.

The input power to the motor of the blower can be calculated as

$$W_{\text{motor,input}} = \frac{Q(P_{\text{out}} - P_{\text{in}})}{\eta_{\text{motor}}\eta_{\text{coupling}}\eta_{\text{blower}}} \tag{2.34}$$

where $W_{\text{fan,input}}$ is the input electric power to the motor of the blower in kW; P_{in} is the static pressure at the inlet of the blower in kPa; P_{out} is static pressure at the outlet of the blower in kPa; Q is the volume flow rate of air in m^3/s; and η_{motor}, η_{coupling}, and η_{fan} are the efficiencies of the motor, coupling and blower, respectively. The input electric power to the motor of a blower can be reduced by selecting high-efficiency motor, coupling, and blower. The operating efficiency of the motor and blower also varies with the change in air flow rate. The characteristic curves of the motor and blower should be analyzed carefully to ensure that the components will be running at the best efficiency region. (P_{out} – P_{in}) represents the differential pressure generated by the blower to overcome the pressure losses of the ducting system. The pressure drops across the filter, cooling coil, and duct are proportional to the square of the velocity of flowing air. The filter and cooling coil of a higher face area and the duct of a bigger cross-sectional area reduce the airflow velocity and contribute to the power consumption of the blower system. However, the capital cost of the filter, cooling coil, and duct will be higher. The ducting system and components should be selected based on the life cycle cost.

2.12 Chiller Plant with Thermal Energy Storage (TES) Tank

The efficiency of a chiller varies with its loading. The optimum loading and corresponding best operating efficiency of a chiller depend on several factors, including compressor design, refrigerant type, heat exchanger design for the evaporator and condenser, and control strategies. Because of the dynamic

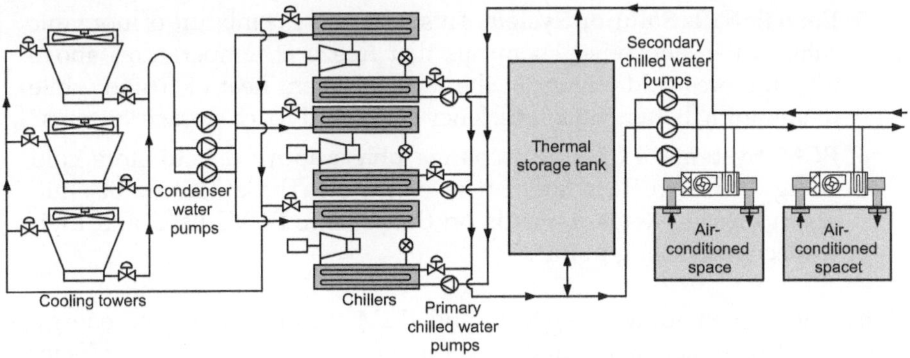

FIGURE 2.32
Thermal energy storage tanks integrated into air-conditioning systems.

nature of building cooling loads, it is not practically possible to operate the chillers at the best efficiency throughout the operational hours. To address this, thermal energy storage (TES) tanks are integrated into air-conditioning systems as shown in Figure 2.32. These tanks allow chillers to consistently operate in the region of optimum loading and best efficiency regardless of the actual cooling demand of the buildings. The TES tanks store excess cold energy during the period of low or no cooling loads of the buildings and utilize the stored cold energy during the period of high cooling loads, ensuring that chillers are operating at optimum loading. The cold energy storage capacity of a tank is expressed in RT-hour. A TES tank with a capacity of 1,000 RT-hours can support a space cooling load of 200 RT for 5 hours.

There are four common types of storage systems designed for thermal energy management: (a) chilled water storage, (b) ice storage, (c) eutectic salt solutions, and (d) phase change material (PCM) systems.

1. **Chilled Water Storage System**: This system stores chilled water at approximately 4–6°C, utilizing its sensible heat as cold energy. A chilled water storage tank with a volume of 550 m^3 can store around 1000 RT-hours of cold energy. Chillers can supply chilled water at 4°C–6°C to these storage tanks without a significant efficiency drop.
2. **Ice Storage System**: Ice storage typically involves encapsulated ice, ice-on-coil, or slurry ice. It relies on the latent heat of fusion of water to store cold energy. Compared to chilled water systems, the ice storage tank required to store 1,000 RT-hours of cold energy is significantly smaller, about 80 m^3, due to the high latent heat of fusion (approximately 333 kJ/kg K) versus the specific heat capacity of water (about 4.2 kJ/kg K). However, chillers must operate at lower temperatures (around −5°C to −10°C) to produce ice, reducing their efficiency by roughly 50%.

3. **Eutectic Salts Solution System**: This system uses a mixture of inorganic salts, water, and other chemicals that freeze at temperatures above 0°C. It stores cold energy in the form of latent heat of fusion while maintaining better chiller efficiency compared to ice storage systems.
4. **PCM Systems**: PCMs also change phase above 0°C to store cold energy through their latent heat of fusion. The efficiency of chillers in these systems depends on the specific PCMs used and their respective freezing points.

Both eutectic salt solution systems and PCM systems store cold energy as latent heat of fusion while allowing for relatively higher chiller efficiency compared to ice storage systems, with specific performance depending on the materials' freezing points.

The feasibility of TES systems is influenced by several factors such as

Electricity Tariffs and Availability: In regions with time-dependent electricity tariffs (peak and off-peak), a TES system can be charged during off-peak hours when the electricity tariff is lower. The stored cold energy is used to meet cooling demands during peak periods, reducing operating costs. Additionally, TES systems can mitigate high electricity demand charges by being charged during low-demand periods. If electricity supply is constrained during the period of high cooling demand, the TES system can be charged when electricity is readily available, providing stored cold energy to support cooling when the supply of electricity is limited.

Cooling Load Profile: For spaces or processes with short and infrequent cooling demand periods, TES systems can effectively manage the load without frequently cycling the chiller on and off or running it inefficiently at low part load. This approach maintains the chiller system's efficiency and reliability.

Operational Continuity and Efficiency: In certain industrial processes, laboratories, and critical operational spaces where cooling interruptions cannot be tolerated, an additional chiller is often run to prevent disruptions caused by chiller failure. This results in the chiller system operating at a low part load with poor efficiency. By using a TES system, the additional chiller can be turned off, improving the loading and efficiency of the operating chiller. If a chiller trips, the TES system can take over the cooling load. Moreover, if a chiller system lacks spare peak capacity but has excess off-peak capacity, the TES system can be charged during off-peak periods to support peak cooling demands, eliminating the need for installing another chiller.

When there is a significant difference between peak and off-peak electricity tariffs and high electricity demand charges, chilled water systems can be utilized to charge TES tanks of larger capacity during off-peak hours. This strategy allows the large TES tank to support the entire cooling load during peak periods, with the chilled water system turned off, a method known as the *full storage strategy*. Alternatively, if the peak cooling demand significantly exceeds the average demand, the cooling load during the peak period can be managed using a *partial storage strategy*. In this approach, the chillers are operated to meet a portion of the peak demand, while the TES tank supplies the remaining cooling load. Evaluating the techno-commercial feasibility of these strategies involves considering the capital costs of TES tanks, chillers, and control systems, as discussed in Chapters 3 and 6. Overall, TES systems enhance operational flexibility, improve energy efficiency, and reduce costs by optimizing the chiller system and leveraging favorable electricity tariffs and availability. A larger-scale application of the TES is known as a district cooling system. Several high-capacity chillers and TES tanks are installed in the district cooling systems.

2.13 Conclusions

This chapter describes the working principles and theoretical fundamentals for evaluating the performance of conventional central air-conditioning chiller plants. The energy performance enhancement opportunities of air-conditioning chiller plants by adopting several technologies such as cascading, multistaging, resetting of chilled water and condenser water temperatures, and sizing and configuring of chillers are illustrated with relevant examples. The achievable energy savings by adopting the genetic algorithm-based optimization strategies for sequencing and energy-smart loading compared to conventional control strategies of chillers are explored. This chapter elaborates on how pumps are installed in parallel and series configurations in primary and secondary pumping systems and explores the interaction between pump curves and system curves under different operating conditions. Additionally, it covers control strategies for the energy-efficient operation of pumping systems and potential energy savings, supported by a case study on the calculation steps, sizing, designing, and control of a condenser water pumping system. This chapter also describes heat and mass transfer processes, along with sizing and control strategies for cooling tower systems under dynamic heat loads and varying ambient conditions. Finally, it highlights the retrofitting opportunities, benefits, and limitations of four common types of TES systems, namely, chilled water storage, ice storage, eutectic salt solutions, and PCM systems. Overall, this chapter provides an overview of the working principles, sizing, and control strategies of central air-conditioning chiller systems, essential for understanding the efforts involved in developing and optimizing the operation of district cooling systems.

References

1. Jayamaha, L. (2006). *Energy-efficient building systems* (1st ed.). McGraw-Hill.
2. Wang, S. K. (2001). *Handbook of air conditioning and refrigeration* (2nd ed.). McGraw-Hill.
3. Wang, S.W., & Ma, Z. J. (2010). Control strategies for variable speed pumps in super high-rise building. *ASHRAE Journal*, 52(7), 36–43.
4. Cengel Y. A., Boles M. A., & Kanoglu, M. (2020). *Thermodynamics an engineering approach* (9th ed.). McGraw-Hill Education.
5. Islam, M. R. (2016). Air-conditioning and mechanical ventilation systems. https://www.ies.org.sg/Tenant/C0000005/PDF%20File/Registry/SCEM/ACMV(1).pdf
6. ASHRAE Handbook. (2001). *Fundamentals volume.* American Society of Heating, Refrigerating and Air-Conditioning Engineers, Inc.
7. McQuiston, F. C., Parker, J. D., & Spitler, J. D. (2005). *Heating, ventilating, and air conditioning analysis and design* (6th ed.). John Wiley & Sons, Inc.
8. Jayamaha, L. (2018). Motor driven systems: Reference manual. https://www.nea.gov.sg/docs/default-source/our-services/energy-efficiency/industrial-sector/scem/mds.pdf
9. Wijeysundera, N. E. (2016). *Principles of heating, ventilating and air conditioning with worked examples* (1st ed.). World Scientific.

3

District Cooling System Design

3.1 Introduction

The increasing worry about climate change and the global commitment to environmental protection necessitate a shift toward renewable technologies, carbon capture utilization and storage (CCUS) technology, and improvements in energy efficiency. Within the current energy disclosure, cooling energy production emerges as a significant contributor to energy consumption. Specifically, approximately 10% of electric energy consumption is attributed to air conditioning or space cooling applications in buildings [1]. The International Energy Agency (IEA) anticipates that space cooling will rank among the most electricity-intensive activities in the coming three decades. This underscores the essential role of cooling energy in our daily lives within modern society.

Various cooling technologies have been developed and deployed globally to cater to diverse needs. Among these, these electrically driven vapor compression-based cooling system stands out due to the energy-efficient design, technical maturity, and reliable operations. The versatility of the vapor compression system, offering a wide capacity range and sustainable refrigerant options, positions itself as the primary choice of cooling applications, surpassing alternatives such as evaporative cooling [2], absorption theory-based cooling technology [3], solid desiccant-based cooling technology [4], and free cooling resources [5].

However, relying solely on vapor compression and conventional cooling technologies prove insufficient, prompting continuous exploration of effective cooling methods with minimal environmental impact but improved commercial impact. The DCS [6–8] has emerged as a sustainable cooling solution offering numerous advantages. These benefits encompass (i) financial savings in terms of capital, operations and maintenance, (ii) enhanced reliability, (iii) flexibility, (iv) scalability, and (v) environmental friendliness. A detailed discussion of these benefits will be presented in subsequent sections of this chapter. Consequently, extensive research analyses and practical applications of DCS have been conducted.

DOI: 10.1201/9781032670164-3

3.1.1 Research Activities

Over the past decade, there has been a significant surge in research interest surrounding DCS technology. Rezaie et al. [9] and Werner [10] conducted thorough reviews, delving into the technical and environmental aspects of district-level heating and cooling systems. Notably, their emphasis was primarily on district heating, particularly in the European region, where a substantial demand for heating energy exists.

In efforts to decrease reliance on fossil fuel-driven electric energy for district cooling and heating applications, analyses exploring the integration of renewable energy technologies, alongside other energy-efficient technologies, with DCS have been undertaken. Additionally, DCS exhibits an economy of scale when compared to individual chiller systems, providing opportunities for integration with various technologies in a commercially viable and cost-effective manner [11, 12]. Gang and his team [11] have explored applications involving renewable resources such as geothermal energy [13], surface water cooling, biomass energy, solar energy, and waste energy coupled with DCS, with a focus on understanding the system at the macro level. Furthermore, they briefly discuss system-level design and optimization. Another study investigates the implementation of a solar thermally driven absorption chiller within the DCS system using the TRNSYS simulation environment [14], specifically in Saudi Arabia, due to its geographic location with high solar irradiance. In addition, the district energy systems powered by nuclear technologies have been studied [15].

Collectively, these studies highlight that DCS, when combined with renewable energy sources, can pave the way for a sustainable cooling solution [16, 17], contingent upon the geographic availability of resources under different operation strategies.

Essentially, the creation of a DCS system hinges on technical feasibility, commercial viability, and the resources at hand. Typically, the core DCS technologies—specifically, the cooling system based on vapor compression technology and the thermal energy storage (TES) tank—constitute pivotal elements in designing a sustainable DCS. Concurrently, renewable energy sources and nuclear resources serve as supplementary yet indispensable components for enhancing environmental sustainability. In other words, all renewable energy resources contribute as sustainable energy inputs to the DCS, with fundamental DCS technologies serving as foundational pillars. In scenarios where economically bustling cities face constraints such as land scarcity and limited natural resources, DCS offers a sustainable cooling solution. By utilizing DCS cooling services, free spaces on building rooftops are created. This facilitates the installation of solar photovoltaic (PV) systems, making solar PV technology the preferred renewable energy integration in congested urban areas. Consequently, the harvested solar electric energy can be injected into the electrical grid or used within the building.

This innovation approach allows us to design a DCS independently of renewable energy components.

3.1.2 Objective

Given the significant investments involved in DCS implementation, the design and optimization of cooling systems are crucial for the deployment of a cost-effective system. This chapter introduces a practical design tool intended for use during the design and planning stages [18] while enhancing analyses in a more detailed manner. The proposed model provides optimal specifications for chiller cooling capacity and TES design.

The initial step involves developing a mathematical model for the DCS, wherein an objective function is established. This function is a straightforward cost metric that encompasses capital expenditure (CAPEX) and operation and maintenance expenses (OPEX) throughout the economic life of the DCS plant. The optimization technique employed is the point interpolation technique. Various case studies are presented with detailed discussions to illustrate the application of the proposed model. The objective of this chapter is to contribute a valuable design tool for optimizing DCS in both industrial and academic settings.

3.2 District Cooling System Design

This chapter proposes a methodology that facilitates optimal DCS design by taking into account technical parameters as well as investment and operational expenses.

As a reminder, a DCS plant, as illustrated in Figure 3.1, constitutes a multi-chiller system. It encompasses numerous chillers, chilled water pumps linked to the chillers or primary chilled water pumps (CP_a), chilled water distribution pumps or secondary chilled water pumps (CP_b), condenser water pumps (CDP), cooling towers (CT), and TES. This configuration is chosen for its notable energy efficiency, environmental friendliness, and economic benefits.

The optimal DCS system design is formulated with a central focus on chiller systems and TES as key design elements. Initially, the cooling requirements of diverse buildings are gathered, and a consolidated local cooling load profile is developed for a typical day for a DCS plant. This is imperative to use the cooling load profile of the hottest day. Subsequently, the design optimization process involving cost minimization over the economic lifespan is conducted to ascertain the optimal cooling energy generation capacity during peak/off-peak hours and the TES capacity, as depicted in Figure 3.2, where

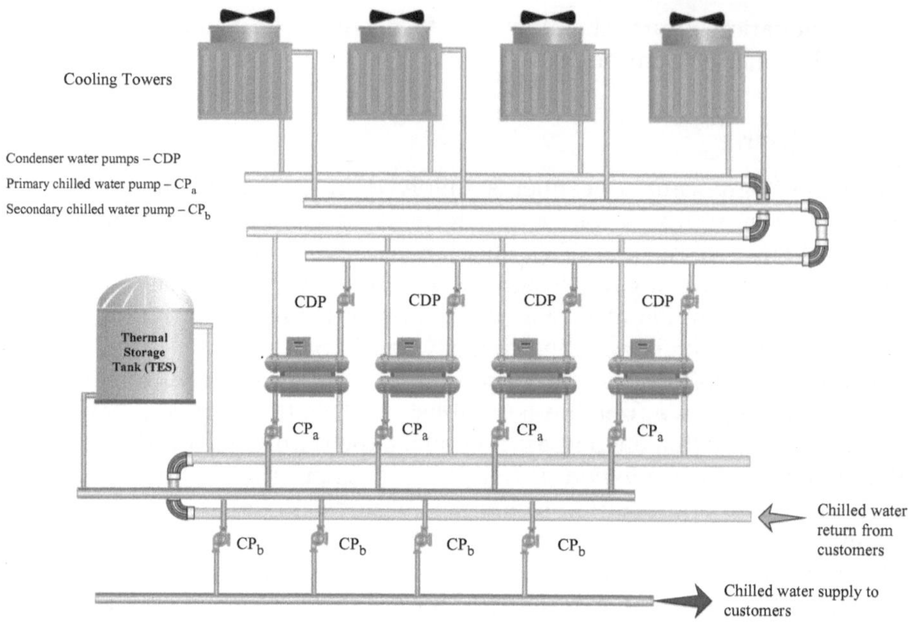

FIGURE 3.1
A typical DCS system with chillers, condenser water pumps, cooling towers, and primary and secondary chilled water pumps.

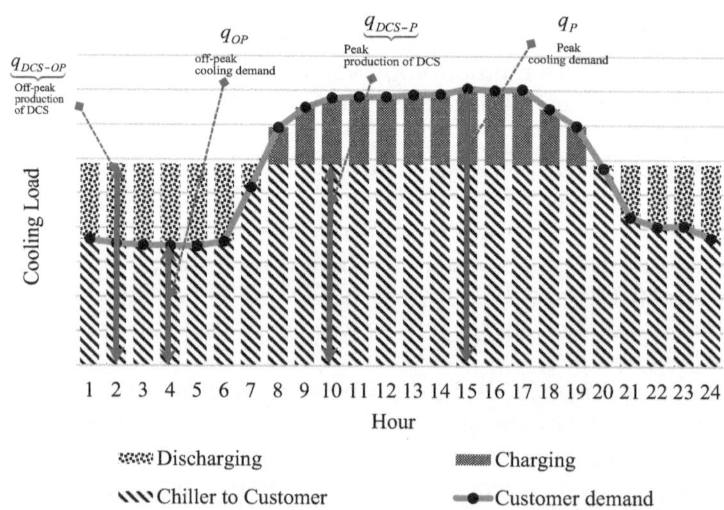

FIGURE 3.2
Hourly load profile of chiller operation with TES charging and discharging [18].

q_{OP} refers to hourly cooling energy demand of the DCS at off-peak hours, q_P refers to hourly cooling energy demand at peak hours, and q_{DCS-OP} and q_{DCS-P} are hourly cooling energy generation of the DCS. The optimization model incorporates local operational and financial cost structures, as well as space constraints. In the final phase, the optimized DCS system is scrutinized using financial metrics in Chapter 4 and compared against a conventional centralized chiller system design.

3.2.1 DCS Cooling Demand and Operation Strategy

First, the total cooling demand of the DCS can be obtained by combining the cooling demand of developments in the district, as shown in Figure 3.3.

The total cooling energy generation of the DCS is then defined as

$$Q_P = \sum_{i=1}^{n} q_{P,i}, \quad Q_{OP} = \sum_{j=1}^{24-n} q_{OP,j} \quad \text{and} \quad Q_{Total} = Q_P + Q_{OP} \tag{3.1}$$

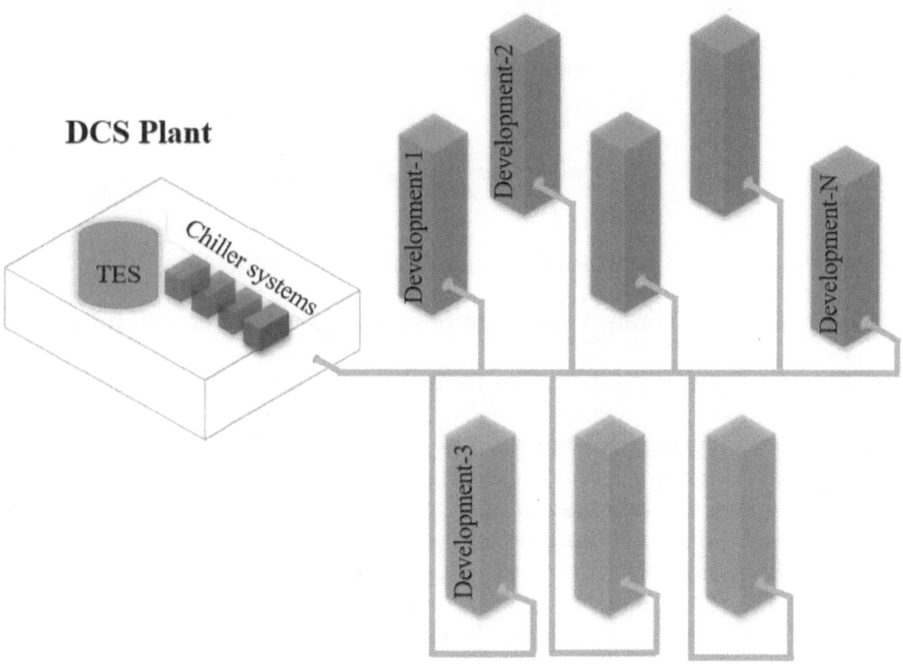

FIGURE 3.3
A simple diagram of a district cooling plant comprising chiller systems and thermal energy storage (TES), supplying chilled water to developments in the district through an underground pipe network.

where Q_P is the total peak hour cooling demand in refrigeration ton hour (RTh), $q_{P,i}$ is hourly peak hour cooling demand in refrigeration ton (RT) at *i*th hour, Q_{OP} is total off-peak cooling demand in refrigeration ton hour (RTh), $q_{OP,j}$ is hourly off-peak cooling demand in (RT) at *j*th hour, and Q_{Total} is total customer cooling demand of a day for a DCS.

The operation of the DCS usually varies between peak and off-peak hours. In practice, off-peak hours will have a lower electricity tariff. To achieve the total customer load Q_{Total}, the DCS system's operating approach is established as

$$Q_{Total} = \left(q_{DCS-OP} \times h_{OP}\right) + \left(q_{DCS-P} \times h_P\right) \tag{3.2}$$

where q_{DCS-OP} and q_{DCS-P} are hourly off-peak and peak hour generation of DCS, h_{OP} is total off-peak hours, and h_P is total peak hours of a day. During off-peak hours, the DCS produces $q_{DCS-OP} > q_{OP}$, while during peak hours, it generates $q_{DCS-P} < q_P$, see Figure 3.2. The chiller capacity installed in the DCS is then specified as

$$q_{DCS-installed} = max\left(q_{DCS-OP}, q_{DCS-P}\right). \tag{3.3}$$

TES is an integral component of DCS and can be configured as

$$Q_{TES} = \left(\left(q_{DCS-OP} \times h_{OP}\right) - \sum_{j=1}^{24-n} q_{OP,j} \right) + \sum_{k=1}^{m} q_{c,k}, \tag{3.4}$$

where $\left(q_{DCS-OP} \times h_{OP}\right)$ represents the cooling energy generated by DCS during off-peak hours while $\sum_{j=1}^{24-n} q_{OP,j}$ signifies the customers' demand for cooling those same off-peak periods. The term $\sum_{k=1}^{m} q_{c,k}$ accounts for the charging of cooling energy during peak hours. Once office hours end, the cooling demand for the DCS typically declines, resulting in less efficient chiller performance. To mitigate this, the TES system becomes crucial by storing surplus cooling energy as in Eq (3.4), allowing the DCS to maintain efficient operations even when the demand is low. The term $q_{c,k}$ is given as

$$q_{c,k} = \begin{cases} 0, if\ q_{P,k} > q_{DCS-P} \\ q_{DCS-P} - q_{P,k}, if\ q_{P,k} < q_{DCS-P} \end{cases}. \tag{3.5}$$

As previously noted, the DCS operates with constant cooling energy output by parameters q_{DCS-OP} and q_{DCS-P}, which stay consistent throughout peak and off-peak hours. This stability ensures the chillers function at maximum efficiency, though adjusting these parameters could further improve system performance.

3.2.2 Optimal Design Procedure

Effective DCS design requires careful evaluation of three critical factors: (i) operational costs or the so-called operating expenses, (ii) initial capital investment/expenditure (CAPEX), and (iii) long-term maintenance expenses throughout the system's lifespan, $\mathfrak{N}$.

Operating Expenses: Operating expenses for energy utilization over the plant's entire life need to be accessed, as given in Eq (3.6).

$$C_{OPEX-1} = \sum_{t=1}^{\mathfrak{N}} C_{OPEX-1,t} \tag{3.6}$$

where the term $C_{OPEX-1,t}$ is annual energy utilization expenses and defined as

$$\begin{aligned} C_{OPEX-1,t} = [&\left(TF_{OP,t} \times \eta_{DCS,t} \times h_{OP} \times q_{DCS-OP} \times 365\right) + \left(TF_{P,t} \times \eta_{DCS,t} \times h_P \times q_{DCS-P} \times 365\right) \\ &+ \left(q_{DCS-installed} \times \eta_{DCS,t} \times TF_{CC,t} \times 12\right)]. \end{aligned} \tag{3.7}$$

In Eq. (3.7), the first two terms reflect the annual costs associated with off-peak and peak electricity consumption, while the third term accounts for the yearly cost of contracted electric capacity. The variables $TF_{OP,t}$, $TF_{P,t}$, and $TF_{CC,t}$ refer to the published electricity rates for off-peak usage ($/kWh), peak usage ($/kWh), and the monthly charge based on contracted power capacity ($/kW). These electricity tariffs are determined on an annual basis using the discount cash flow method [19].

$$\begin{aligned} TF_{OP,t} &= TF_{OP,1} \times (1+i)^{t-1} \\ TF_{P,t} &= TF_{P,1} \times (1+i)^{t-1} \\ TF_{CC,t} &= TF_{CC,1} \times (1+i)^{t-1} \end{aligned} \tag{3.8}$$

where i represents the annual escalation rate or discounted rate in percentage terms, "t" denotes a specific year, $TF_{OP,1}$ refers to the off-peak electricity tariff in the starting year, $TF_{P,1}$ is peak electric energy tariff at initial year, and $TF_{CC,1}$ corresponds to the contracted capacity charges in the first year. Similarly, the efficiency of the DCS plant is estimated while considering the degradation of the chiller plant

$$\eta_{DCS,t} = \eta_{DCS,1} \times (1+d)^{t-1} \tag{3.9}$$

where $\eta_{DCS,1}$ is the efficiency of the plant at initial year in kWh/RTh and d is the annual degradation rate of the chiller plant in percentage.

CAPEX: The initial capital expenditure, CAPEX, is defined as

$$C_{CAPEX} = \left(q_{DCS-installed} \times U_{chiller-system}\right) + \left(Q_{TES} \times U_{TES}\right) + C_{Pipe-network} \tag{3.10}$$

where $U_{chiller-system}$ represents the cost per unit of the chiller system in \$/RT, U_{TES} indicates the cost per unit of TES system in \$/RTh, and $C_{Pipe-network}$ represents the cost of underground pipe network in dollars. These costs are incurred only once for the development of the DCS plant. However, the pipe network cost is excluded from this analysis due to its variability, which depends on both the distance from the DCS plant and the number of customers it serves. Nonetheless, it is essential to recognize that in practical applications, these costs must be taken into account.

Maintenance Cost: The maintenance cost is also considered a form of operating expenses, as it involves the ongoing expenditure required to keep the DCS plant functioning efficiently and effectively throughout its economic life. The overall maintenance cost is defined as

$$C_{OPEX-2} = \sum_{t=1}^{\mathfrak{n}} C_{Mtn.t}. \tag{3.11}$$

In Eq (3.11), $C_{Mtn,t}$ is annual maintenance cost encompassing routine inspections, repairs, and replacements of components, given as

$$C_{Mtn,t} = \left(q_{DCS-installed} \times U_{Mtn,t}\right) \tag{3.12}$$

where $U_{Mtn,t}$ represents the annual maintenance cost per unit of the chiller system in \$/RT for "t" year and

$$U_{Mtn,t} = U_{Mtn,1} \times (1+i)^{t-1}. \tag{3.13}$$

Objective Function for Design Optimization: In order to ensure commercial viability by optimizing both the technical design and the associated investment and operational costs, the objective function is formulated as

$$f\left(q_{DCS-P}, q_{DCS-OP}\right) = \bar{C}_{OPEX-1} + \bar{C}_{CAPEX} + \bar{C}_{OPEX-2} \tag{3.14}$$

The variables $\bar{C}_{OPEX-1}$, $\bar{C}_{CAPEX}$, and $\bar{C}_{OPEX-2}$ represent the normalized costs, defined as

$$\bar{C}_{OPEX-1} = C_{OPEX-1} / \tilde{C}_{OPEX-1}, \quad \bar{C}_{CAPEX} = C_{CAPEX} / \tilde{C}_{CAPEX}, \text{ and}$$
$$\bar{C}_{OPEX-2} = C_{OPEX-2} / \tilde{C}_{OPEX-2}. \tag{3.15}$$

Normalization is performed based on the respective costs of the DCS plant—OPEX ($\tilde{C}_{OPEX-1}$), CAPEX ($\tilde{C}_{CAPEX}$), and maintenance cost ($\tilde{C}_{OPEX-2}$)—assuming steady operating conditions typically for the DCS systems ($q_{DCS-installed} = q_{DCS-OP} = q_{DCS-P}$). This implies that the operating conditions are treated as the same during both peak and off-peak periods. By normalizing, the OPEX and CAPEX components are balanced, ensuring they are equally weighted in the optimization process.

$$\mathbf{q}^* = \arg\left(\min f\left(q_{DCS-P}, q_{DCS-OP}\right)\right), \tag{3.16}$$

where $\mathbf{q}^*$ represents an optimal DCS design that specifies the chiller capacity during both peak and off-peak periods. It is determined by minimizing the objective function using the Karush-Kuhn-Tucker (KKT) method [20, 21], with constraints given in Eqs. (3.2) and (3.4). Using these optimized design parameters, the ideal chiller capacity and TES for the DCS can be defined. The optimization process is illustrated in Figure 3.4. In summary, input parameters such as peak/off-peak periods, local electricity tariffs, annual efficiency degradation, discount rate and the costs of chillers, TES, and maintenance are used in Eqs. (3.1)–(3.13). The objective function in Eq. (3.14) is then constructed for optimization, with initial values for q_{DCS-OP} and q_{DCS-P}. The process continues until an optimal DCS design of q_{DCS-P}, q_{DCS-OP} and Q_{TES} is achieved, following Eq. (3.14). The resulting design is subsequently verified through life cycle cost analysis.

3.3 Case Studies

3.3.1 Problem Definition

A case study [18] involving a district with ten buildings is undertaken to apply the algorithm described in Section 3.2. Singapore is selected as the study location due to the availability of industry data and its relevance as a representative tropical environment where DCS technology is widely implemented. Table 3.1 presents the cooling requirements for each building, which are estimated by simulating cooling demand patterns typically of Singapore's office buildings, mixed-use developments, and data centers. This dataset forms the basis for developing an optimal DCS design using the proposed methodology.

The **optimal DCS** is designed, and it is subsequently contrasted with various potential DCS designs, stemming from DCS expertise and potentially encompassing a wide spectrum of DCS configurations. Different DCS cases are studied, utilizing three different variables: (i) three distinct electric energy

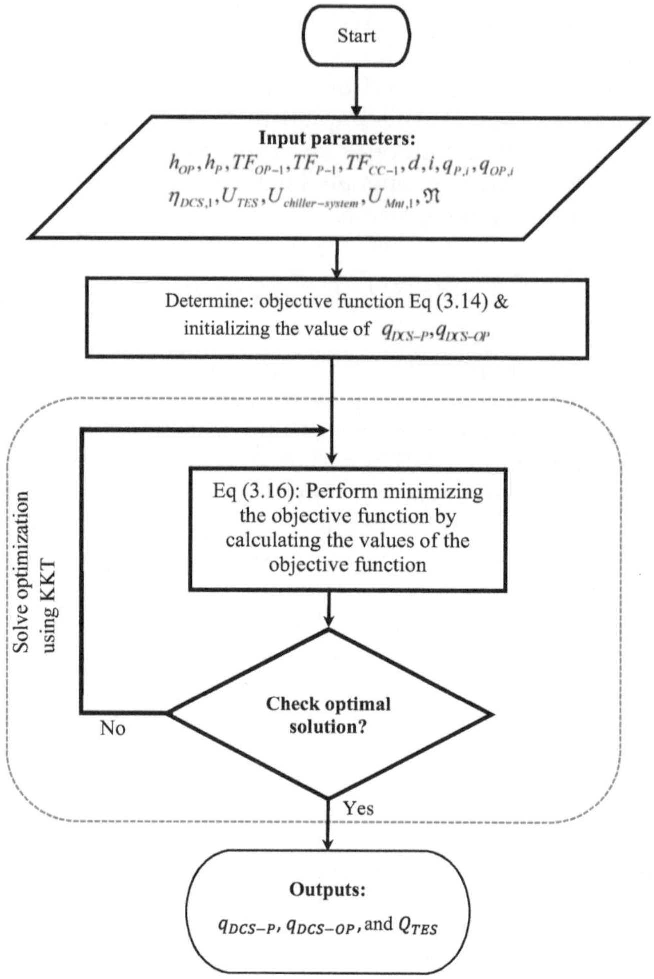

FIGURE 3.4
The optimization flow chart for DCS design [18].

tariff structures, (ii) three different TES capacities, and (iii) three different CAPEX structures. This results in the examination of 27 DCS cases, as given in Table 3.2. Certainly, the cases are not exhaustive, but they are adequately comprehensive for this analysis.

For the analyses, the three distinct thermal storage capacities are denoted as (a) limited thermal energy storage (LTES), (b) optimal thermal energy storage (OTES), and (c) excess thermal energy storage (ETES). The three diverse CAPEX structures are identified as: (A) utilizing present investment rates of chiller systems and TES based on market intelligence, (B) incorporating

TABLE 3.1
Cooling Demand for Ten Buildings in Refrigeration Ton (RT)

Hour	Building Cooling Demand Load (RT)[b]										Total Hourly Demand Load (RT)
	B1[a]	B2	B3	B4	B5	B6	B7	B8	B9	B10	
1	196	1,000	0	368	417	140	429	489	288	342	3,670
2	187	1,000	0	351	397	140	404	463	276	328	3,545
3	184	1,000	0	344	390	140	395	453	271	322	3,499
4	182	1,000	0	342	387	140	391	449	270	320	3,480
5	182	1,000	0	341	386	140	390	448	269	320	3,476
6	191	1,000	0	359	407	140	416	475	282	335	3,604
7	254	1,000	300	525	540	350	589	655	445	529	5,188
8	346	1,000	600	563	735	525	841	919	634	753	6,915
9	384	1,000	600	600	850	560	945	1,028	698	829	7,495
10	384	1,000	600	638	850	595	1,056	1,104	739	808	7,774
11	391	1,000	600	675	850	630	1,020	1,087	748	808	7,808
12	398	1,000	600	750	850	700	1,094	1,144	797	800	8,132
13	400	1,000	600	675	850	700	990	1,075	773	808	7,871
14	395	1,000	600	675	850	630	1,031	1,098	753	855	7,887
15	393	1,000	600	713	850	665	1,047	1,106	769	913	8,055
16	382	1,000	600	720	850	672	1,018	1,076	757	922	7,997
17	382	1,000	600	720	850	672	1,030	1,084	761	931	8,030
18	371	1,000	600	600	850	560	909	991	681	903	7,464
19	336	1,000	600	623	715	581	815	892	643	764	6,970
20	280	1,000	300	525	595	490	660	730	533	633	5,746
21	244	1,000	0	458	519	140	562	627	352	418	4,321
22	224	1,000	0	421	477	140	507	570	326	387	4,052
23	226	1,000	0	424	480	140	511	575	328	389	4,073
24	200	1,000	0	375	425	140	441	501	294	349	3,724

Notes:
[a] B1 refers to Building 1.
[b] Thanks to Keppel DHCS Pte Ltd for the contributions in estimating the cooling demand in a district cooling zone.

25% higher investment rates than the present rates, and (C) implementing 50% higher investment rates than the present rates. Similarly, the three electric energy tariff variations are specified as follows: (1) off-peak tariff set at 30% of the peak tariff, (2) off-peak tariff set at 60% of the peak tariff, and (3) off-peak tariff set at 90% of the peak tariff. As given in Table 3.2, the case "LTES-A-1" represents a DCS analysis with limited TES capacity, utilizing present investment rates for chiller systems and TES sourced from market intelligence and employing off-peak tariff set at 30% of the peak tariff.

TABLE 3.2

Different Scenarios for DCS Design Analysis

	Limited TES, LTES 30% Less Storage Compared to OTES (a)[a]			Optimal TES, OTES, Arising from Proposed Optimization Methodology (b)			Excess TES, ETES 30% More Storage Compared to OTES (c)		
	Present investment rates of chillers and TES (A)[b]	25% higher investment rate than the present rates (B)	50% higher investment rates than the present rates (C)	Present investment rates of chillers and TES (A)	25% higher investment rates than the present rates (B)	50% higher investment rates than the present rates (C)	Present investment rates of chillers and TES (A)	25% higher investment rates than the present rates (B)	50% higher investment rates than the present rates (C)
Off-peak tariff = 30% peak tariff (1)[c]	Case: LTES-A-1	LTES -B-1	LTES-C-1	OTES -A-1	OTES -B-1	OTES-C-1	ETES -A-1	ETES -B-1	ETES-C-1
Off-peak tariff = 60% of Peak tariff (2)	LTES-A-2	LTES-B-2	LTES-C-2	OTES-A-2	OTES-B-2	OTES-C-2	ETES-A-2	ETES-B-2	ETES-C-2
Off-peak tariff = 90% of Peak tariff (3)	LTES-A-3	LTES-B-3	LTES-C-3	OTES-A-3	OTES-B-3	OTES-C-3	ETES-A-3	ETES-B-3	ETES-C-3

Notes:

[a] (a), (b), and (c) refer to LTES, OTES, and ETES.

[b] (A), (B), and (C) stand for present investment rate, 25% more investment rate, and 50% more investment rate than the present rate, respectively.

[c] (1), (2), and (3) represent different ratio of peak and off-peak tariffs.

3.4 Results

On examining 27 DCS cases, see Table 3.2, it is imperative to contemplate three different DCS designs, each corresponding to different thermal storage capacities. This consideration arises from the interdependence between the installed chiller capacity and thermal storage capacity in DCS design. Additionally, the other two factors—electric energy tariffs and CAPEX—play a crucial role in assessing the commercial viability of each case.

Load Profile and DCS Design: Consideration of different thermal storage capacities and the practical DCS designs are illustrated in Figures 3.5–3.7 and Table 3.3. Figure 3.6 specifically displays the optimal DCS design derived from the proposed optimization methodology discussed in Section 3.2. The remaining two DCS designs, see Figures 3.5 and 3.7, are also demonstrated to facilitate a comparison with the optimal DCS design.

Figure 3.5 shows a DCS design considering limited TES capacity, incorporating 9,000 RT of chiller capacity ($q_{DCS-installed} = 6,458\,RT$ but rounded to 7,000 RT plus a 2,000 RT backup chiller) and a TES capacity of 15,407 RTh. This configuration maintains a consistent operational load for the DCS, q_{DCS-OP} and q_{DCS-P}, allowing the chillers to function at their most efficient level.

In Figure 3.6, an optimal DCS design derived from the proposed methodology in Section 3.2 is presented. The optimal installed chiller capacity is 8,000 RT ($q_{DCS-installed} = 5,866\,RT$ but rounded to 6,000 RT plus a standby 2,000 RT chiller), and TES capacity is 22,010 RTh.

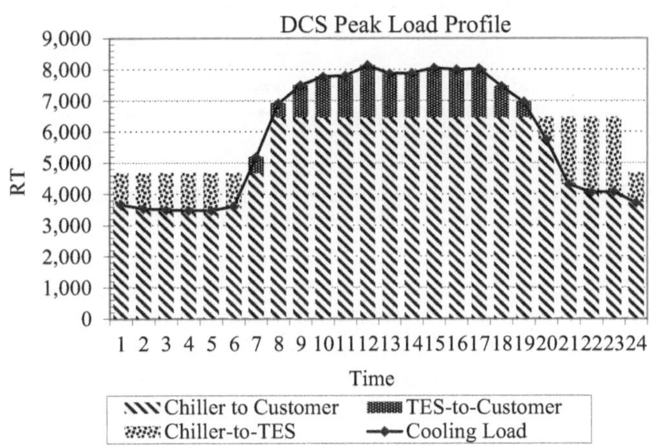

FIGURE 3.5
DCS design with limited thermal storage capacity (LTES; 30% less storage capacity compared to the optimal thermal storage as given in Table 3.2).

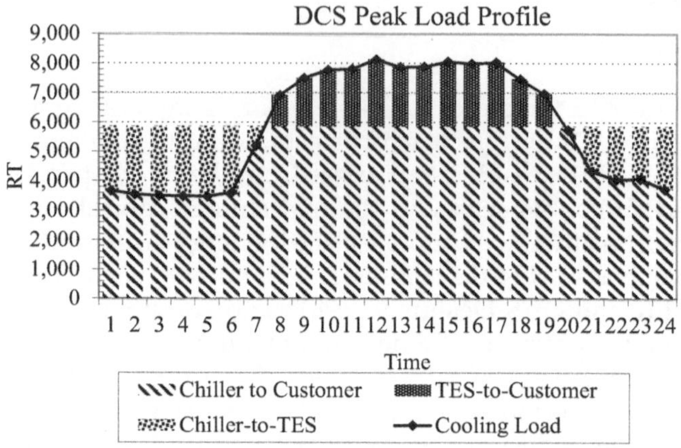

FIGURE 3.6
DCS design with the optimal thermal storage capacity.

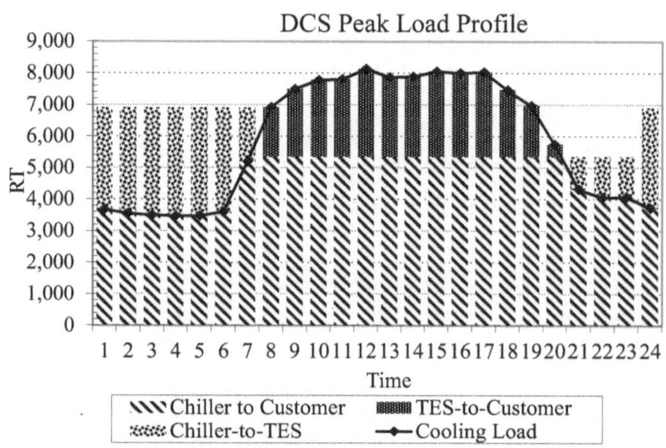

FIGURE 3.7
DCS design with an excess thermal storage capacity (ETES; 30% more storage capacity compared to the optimal thermal storage as given in Table 3.2).

Figure 3.7 showcases a DCS design that considers excess TES capacity. In this analysis, space is not a constraint for DCS developments. The installed chiller capacity for this design is 9,000 RT ($q_{DCS-installed} = 6,900\,RT$ but rounded to 7,000 RT plus a standby 2,000 RT chiller), with a TES capacity of 28,613 RTh.

Standardization of Space Requirement: Following the DCS design, standardizing the spatial requirements of chiller systems and TES becomes essential, even though achieving practical standardization may be challenging.

TABLE 3.3
DCS Designs for Three Different Thermal Energy Storage Capacity

	DCS Design for LTES	Optimal DCS Design	DCS Design for ETES
Installed chiller capacity including standby chiller in RT	9,000 RT (7,000 RT for operation and 2,000 RT for standby capacity)	8,000 RT (6,000 RT for operation and 2,000 RT for standby capacity)	9,000 RT (7,000 RT for operation and 2,000 RT for standby capacity)
Thermal storage capacity in RTh	15,407 RTh (70% of optimal TES)	22,010 RTh (optimal TES)	28,613 RTh (130% of optimal TES)
DCS peak operation load	6,458 RT	5,866 RT	5,349 RT
DCS off-peak operation load	4,681 RT	5,866 RT	6,900 RT
Space requirement (m^2) for both chiller system and TES	979.12 m^2	1014.06 m^2	1207.40 m^2

Nevertheless, this standardization aids in the analysis of all 27 cases concerning both financial and technical considerations.

Drawing upon industry expertise and market insights, the spatial needs for a chiller system that includes a condenser water pump, a chilled water pump, a chiller, and a TES system is established. As demonstrated in Figure 3.8, a chiller system (excluding a cooling tower) is estimated as ~132 m^2 ($W \times L = 6$ m × 22 m). The estimate takes into account the dimensions of a typical 2,000 RT chiller along with its associated pumps. It is essential to note that this figure is provided for analytical purposes within the context of this book. In practical applications, the actual space requirement may vary, influenced by factors such as the chiller's capacity, pump specification, system configuration such as the equipment arrangement and layouts, and safety and accessibility.

In establishing the spatial parameters of TES, a standardized TES height of 25 m is employed. However, it is important to emphasize that this designated TES height is merely a hypothetical value, anticipated to facilitate effective temperature stratification within the tank, and it will vary in reality. For the purpose of the analyses, a consistent height of 25 m is maintained for the tank, while allowing the base area of the tank to vary.

Apart from accommodating the chiller system and TES tank, it is crucial to allocate space for walkways, chiller maintenance, retrofitting activities, and a control room. As a result, an additional 20% has been factored to address the spatial requirement. The DCS designs are given in Table 3.3.

Life Cycle Cost Analysis (LCCA): After the design of DCS, some technical and commercial parameters employed in the analyses are defined in Table 3.4 to conduct the life cycle cost analysis, engaging the formula given in Eq. (3.17).

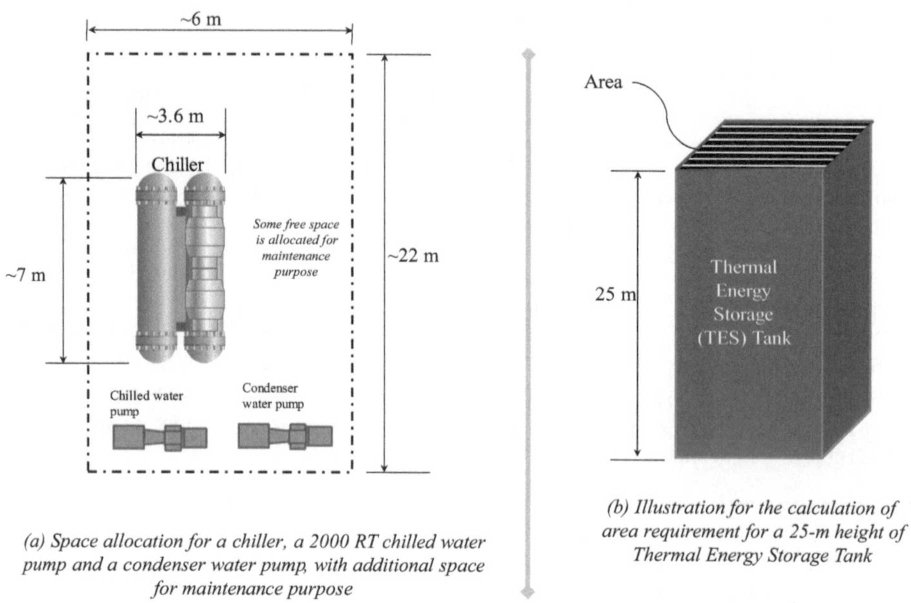

FIGURE 3.8
A typical space requirement for a chiller system: (a) space allocation for a chiller; a chilled water pump and a condenser water pump; (b) a thermal energy storage tank with a height of 25 m.

$$LCCA = C_{CAPEX} + C_{OPEX-1} + C_{OPEX-water} + C_{OPEX-2} + C_{Rental} \tag{3.17}$$

where C_{OPEX-1}, C_{CAPEX}, and C_{OPEX-2} are given in Eqs (3.6), (3.10), and (3.11), respectively. Additional two terms, $C_{OPEX-water}$ and C_{Rental} are introduced

$$C_{OPEX-water} = \sum_{t=1}^{\mathfrak{N}} C_{water,t} = \sum_{t=1}^{\mathfrak{N}} \left(TF_{water,1} \times (1+i)^{t-1} \times W \right)$$

and

$$C_{Rental} = \sum_{t=1}^{\mathfrak{N}} \left(TF_{rental,1} \times (1+i)^{t-1} \times A \times 12 \right) \tag{3.18}$$

where W is the annual water consumption in m^3, $TF_{water,1}$ is water tariff \$/m^3, $TF_{rental,1}$ is monthly rental tariff in \$/m^2 given in Table 3.4, and A is the rental space in m^2.

Tables 3.5–3.7 show the comparison of LCCA for different cases stated in Table 3.2. The result represents that at any investment rate and at any combination of peak and off-peak electric energy tariff, the optimal DCS design always offers the lowest life cycle cost in terms of CAPEX, OPEX, and rental cost, indicating how promising the proposed optimization methodology is.

TABLE 3.4
Input Parameters for DCS Design Studies

Input Parameters		
Off-peak hours	h_{OP}	8 hours
Peak hours	h_P	16 hours (7:00 am to 11:00 pm period)
Peak electric energy tariff	TF_{P-1}	\$0.2491/kWh
Contract capacity tariff [22]	TF_{CC-1}	\$13.44/kW/month
DCS system efficiency	$\eta_{DCS,1}$	0.65 kW/RT or kWh/RTh
Unit cost of thermal energy storage [18]	U_{TES}	\$250/RTh
Unit cost of chiller system[a]	$U_{chiller-system}$	\$3,000/RT
Unit cost for chiller maintenance [23]	$U_{Mnt,1}$	\$50/RT/annum
Unit cost for water [23]	$TF_{water,1}$	\$2.30/m^3
Economic life of chiller system	$\mathfrak{N}$	20 years
Year-on-year efficiency degradation rate	d	0.5%
Annual escalation or discount rate [24]	i	3%[b]
Corporate income tax [25]	i_{tax}	17%
Rental cost [26]	$TF_{rental,1}$	\$32/m^2/month[c]

Notes:
[a] Unit cost of chiller system is based on 2023 data, from industry expertise.
[b] Year-on-year consumer price development in Oct 2023 is of 3.3%, and it is still valid to assume 3%.
[c] Rental cost could vary depending on location and types of buildings. The rental cost of \$32/m^2/month is considered as an average rental cost for industrial estate.
Note: Dollar (\$) throughout this chapter refers to Singapore Dollar.

TABLE 3.5
LCCA Comparison for Nine Cases That Encompass Three Different Energy Storage Capacities and Three Different Capital Expenditure (CAPEX) Structure, While Keeping Electric Energy Tariff During Off-Peak Times at 30% of the Peak Tariff

$TF_{OP} = 30\% \times TF_P$ **(1)**	**Limited TES (a)**	**Optimal TES (b)**	**Excess TES (C)**
Present investment rate (A)	\$290,704,807 *(case: LTES-A-1)*	**\$275,504,731** ***(case: OTES-A-1)***	\$276,869,998 *(case: ETES-A-1)*
25% higher investment rate (B)	\$298,417,744 *(case: LTES-B-1)*	**\$282,880,356** ***(case: OTES-B-1)***	\$285,408,310 *(case: ETES-B-1)*
50% higher investment rate (C)	\$306,130,682 *(case: LTES-C-1)*	**\$290,255,981** ***(case: OTES-C-1)***	\$293,946,623 *(case: ETES-C-1)*

Analyzing the LCCA for various cases will provide insights into (i) the functioning of the proposed optimization system, (ii) the factors contributing to the overall cost, and (iii) the interconnected relationships among these factors. For in-depth discussion purpose, individual cost components are investigated with respect to varying TES capacities and installed chiller

TABLE 3.6

LCCA Comparison for Nine Cases That Encompass Three Different Energy Storage Capacities and Three Different Capital Expenditure (CAPEX) Structure, While Keeping Electric Energy Tariff During Off-Peak Times at 60% of the Peak Tariff

$TF_{OP} = 60\% \times TF_P$ (2)	Limited TES (a)	Optimal TES (b)	Excess TES (C)
Present investment rate (A)	\$309,996,360 *(case: LTES-A-2)*	**\$299,680,652** ***(case: OTES-A-2)***	\$305,308,709 *(case: ETES-A-2)*
25% higher investment rate (B)	\$317,709,297 *(case: LTES-B-2)*	**\$307,056,277** ***(case: OTES-B-2)***	\$313,847,022 *(case: ETES-B-2)*
50% higher investment rate (C)	\$325,422,235 *(case: LTES-C-2)*	**\$314,431,902** ***(case: OTES-C-2)***	\$322,385,334 *(case: ETES-C-2)*

TABLE 3.7

LCCA Comparison for Nine Cases That Encompass Three Different Energy Storage Capacities and Three Different Capital Expenditure (CAPEX) Structure, While Keeping Electric Energy Tariff During Off-Peak Times at 60% of the Peak Tariff

$TF_{OP} = 90\% \times TF_P$ (3)	Limited TES (a)	Optimal TES (b)	Excess TES (C)
Present investment rate (A)	\$328,311,635 *(case: LTES-A-3)*	**\$322,633,116** ***(case: OTES-A-3)***	\$332,308,237 *(case: ETES-A-3)*
25% higher investment rate (B)	\$336,024,573 *(case: LTES-B-3)*	**\$330,008,741** ***(case: OTES-B-3)***	\$340,846,549 *(case: ETES-B-3)*
50% higher investment rate (C)	\$343,737,510 *(case: LTES-C-3)*	**\$337,384,366** ***(case: OTES-C-3)***	\$349,384,862 *(case: ETES-C-3)*

capacities, as given in cases of LTES-A-1, OTES-A-1, and ETES-A-1. These particular cases are selected for analysis due to the considerable contrast between off-peak and peak electric energy prices with the former being only 30% of the latter, see Table 3.5. This substantial disparity in electric energy tariffs between peak and off-peak periods offers an opportunity to demonstrate the notable impact of TES capacity on operational costs.

I. **CAPEX for Chiller System and TES System**: First, CAPEX component is analyzed. Observationally, the CAPEX for the chiller system in the optimal DCS design consistently appears lower compared to the other two designs, as illustrated in Figures 3.5–3.7. This is primarily attributed to the fact that in the optimal DCS design, the hourly production during both peak and off-peak periods is identical, $q_{DCS-OP} = q_{DCS-P}$, in almost every cases. In contrast, for the DCS design with limited TES capacity (LTES-A-1; Figure 3.5), the hourly off-peak production is lower than the hourly peak production, $q_{DCS-OP} < q_{DCS-P}$, necessitating a higher chiller installed capacity. Similarly, in the DCS design featuring excess TES capacity (ETES-A-1; Figure 3.7), the hourly off-peak production surpasses the hourly peak production, $q_{DCS-OP} > q_{DCS-P}$. However, the capacities of TES for limited TES, optimal TES, and Excess TES cases, along with their respective costs, trend similarly, that is, $(Q_{TES})_{LTES-A-1} < (Q_{TES})_{OTES-A-1} < (Q_{TES})_{ETES-A-1}$. These facts collectively contribute to the minimization of CAPEX for the chiller system with TES capacity in the optimized DCS design.

II. **Costs Related to Electricity Consumption**: Second, three important electricity tariffs are considered—(i) monthly registered contracted capacity charges, (ii) peak, and (iii) off-peak electric energy charges—for a detailed analysis of electric energy consumption. The contracted capacity charge pertains to the fees associated with the maximum potential power capacity contracted by consumers, ensuring that electricity providers can accommodate the registered power capacity. Peak and off-peak electric energy tariffs represent the pricing set by electric energy providers, typically with peak tariffs being higher than off-peak tariffs. This is because daytime electricity utilization incurs higher costs due to increased strain on the power grid and associated management challenges for power transmission and distribution networks. In addition, commercial market mechanisms play a crucial role. In the analysis, it is found that the contracted capacity cost of OTES-A-1 is consistently lower than that of LTES-A-1 and ETES-A-1. This discrepancy arises because the DCS plant operation load of OTES-A-1 is low, as illustrated in Figure 3.6. The DCS operation load of LTES-A-1 is high during peak hours to fulfill the peak hour cooling demand requirement, as in Figure 3.5. Conversely, the operation load of ETES-A-1 is high during off-peak periods to produce more cooling energy for a larger TES capacity, as illustrated in Figure 3.7.

III. **Maintenance Cost**: Third, another operational expenditure (OPEX) component, maintenance cost, is examined. Maintenance involves various activities, including but not limited to regular inspection, cleaning of chiller components, pressure system management, refrigerant management, electrical system inspection, and water treatment system inspection. However, TES system typically incurs minimal to negligible maintenance costs over its economic lifespan. Consequently, higher installed chiller capacity results in increased maintenance costs. This suggests that OTES-A-1 represents a lower lifecycle maintenance cost compared to LTES-A-1 and ETES-A-2. The installed capacities of OTES, LTES, and ETES can be found in Table 3.3.

IV. **Rental Cost and Water Consumption Cost**: Fourth, the cost related to the occupied space is monitored. The space requirement of LTES is lowest among all the cases, leading to a lowest lifecycle rental cost. It is mainly due to the less space requirement of TES in the LTES case. Moving on, the water consumption cost remains consistent across all cases. This is because in all DCS designs, the total cooling energy production and heat rejection remain constant over the economic lifespan (water consumption in DCS operation occurs during the heat rejection process of chiller systems through evaporation means, with further details available in Chapter 2).

Having considered the diverse impacts of the DCS designs, Figure 3.9 illustrates various unit cost components measured \$/RTh. It reveals that installed chiller capacity, installed TES capacity, maintenance, electricity cost, and rental costs are intricately interconnected. The analysis underscores the significance of off-peak electric energy cost, which stands at only 30% of the peak electric energy cost. Despite this substantial difference in electric energy tariffs between peak and off-peak periods, it is observed that excess TES capacity does not necessarily result in the best lifecycle cost. This suggests that the proposed DCS design methodology effectively showcases the optimization potential of both CAPEX and operational expenses, ultimately maximizing asset utilization.

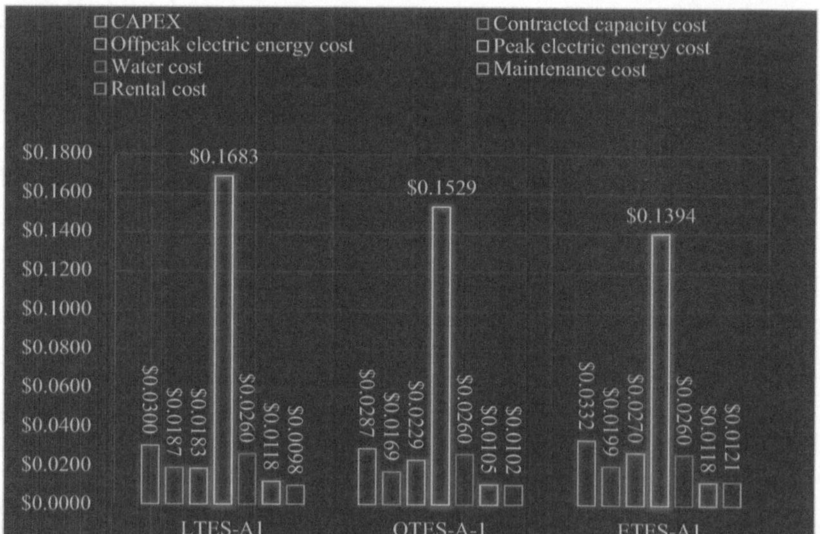

FIGURE 3.9
The unit cost factors of the DCS designing \$/RTh, including CAPEX, electric energy cost, water cost, rental cost, and maintenance cost.

3.5 Conclusions and Discussion

DCS is a practical sustainable cooling solution for busy and highly populated business cities all over the world. DCS provides promising energy and cost savings and environmental friendliness and creates commercial opportunities. Since DCS creates invaluable benefits to the community in a sustainable manner, it increasingly becomes a well-known cooling solution. The detailed design concept, potential design optimization, and a complete understanding are key drivers in the decision-making process of DCS developments.

To recall, Chapter 2 provides essential DCS components to a complete DCS system. Theories behind chillers, cooling tower, pumps, and TES tank are discussed thoroughly. Practical issues and potential solutions are provided. Chapter 2 serves as a typical and fundamental understanding of DCS. With the foundation of DCS delivered in Chapter 2, Chapter 3 provides the detailed optimization model for an optimal DCS design. The objective function is a cost minimization function that accounts for technical design, initial investment, and operational expenses over the economic lifespan. Design case studies have been conducted for different design scenarios covering most potential DCS system configurations. From the analyses, it is found that:

1. The proposed methodology contributes a useful and promising technical framework that should be studied together with financial analyses. Note that the financial analyses will be discussed in Chapter 4.
2. Cost comparison of different design scenarios is presented. The result points out that the optimal DCS design offers the best cost savings.
3. The lifecycle cost analysis implies the impact of initial investment CAPEX and the operating expenses OPEX to the optimal DCS design. The proposed methodology facilitates the DCS industry, particularly the DCS developers, for a commercially sound decision-making process during the DCS development stage.

Acknowledgments

The authors would like to thank Keppel DHCS Pte. Ltd. for the contribution of data and discussions during the course of the work.

References

1. https://www.iea.org/reports/the-future-of-cooling
2. Kapilan, N., Isloor, A.M., & Karinka, S. (2023). A comprehensive review on evaporative cooling systems. *Results in Engineering*, 18, 101059. https://doi.org/10.1016/j.rineng.2023.101059
3. Nikbakhti, R., Wang, X., Hussein, A.K., Iranmanesh, A. (2020). Absorption cooling systems – Review of various techniques for energy performance enhancement. *Alexandria Engineering Journal*, 59, 707–738. https://doi.org/10.1016/j.aej.2020.01.036
4. Sahlot, M., & Riffat, S. B. (2016, December 15). Desiccant cooling systems: A review. *International Journal of Low Carbon Technologies*, 11(4), 489–505. https://doi.org/10.1093/ijlct/ctv032
5. Ding, J., Zhang, H., Leng, D., Xu, H., Tian, C., Zhai, Z. (2022). Experimental investigation and application analysis of free cooling and heat recovery for data centres. *International Journal of Refrigeration*, 136, 142–151. https://doi.org/10.1016/j.ijrefrig.2022.01.003
6. Phetteplace, G., Abdullah, S., Anderpont, J., Bahnfleth, D., Ghani, A.A., Meyer, V., & Tredinnick, A. (2013). *District cooling guide*. American Society of Heating, Refrigerating, and Air-Conditioning Engineers (ASHRAE). ISBN 978-1-936504-42-8.
7. https://www.districtenergy.org/topics/district-cooling
8. Shumoda, Y., Nagota, T., Isayama, N., & Mizuno, M. (2008). Verification of energy efficiency of district heating and cooling system by simulation considering design and operation parameters. *Building and Environment*, 43, 569–577.
9. Rezaie, B., Lake, A., & Beyerlein, S. (2016). Review of district heating and cooling systems for a sustainable future. *Renewable and Sustainable Energy Reviews*, 67, 417–425.
10. Werner, S. (2017). International review of district heating and cooling. *Energy*, 137, 617–631.
11. Gang, W., Wang, S., & Xiao, F.. Gao, D. C. (2016). District cooling systems: Technology integration, system optimisation, challenges and opportunities for applications. *Renewable and Sustainable Energy Reviews*, 53, 253–264.
12. Khir, R., & Mohamed, H. (2015). Optimisation models for a single-plant district cooling system. *European Journal of Operational Research*, 247, 648–658.
13. Carotenuto, A., Figaj, R. D., & Vanoli, L. (2017). A novel solar-geothermal district heating, cooling and domestic hot water system: Dynamic simulation and energy economic analysis. *Energy*, 141, 2652–2669.
14. Franchini, G., Brumana, G., & Perdichizzi, A. (2018). Performance prediction of a solar cooling system in Riyadh, Saudi Arabia – A case study. *Energy Conversion and Management*, 166, 372–384.
15. Abushaman, H. A. S., & Skoda, R. (2022). Nuclear energy for district cooling systems – Novel approach and its eco-environmental assessment method. *Energy*, 250, 123824.

16. Inayat, A., & Raza, M. (2019). District cooling system via renewable energy sources: A review. *Renewable and Sustainable Energy Reviews*, 107, 360–373. https://doi.org/10.1016/j.rser.2019.03.023
17. Sameti, M., & Haghighat, F. (2019). Hybrid solar and heat-driven district cooling system: optimal integration and control Strategy. *Solar Energy*, 183, 260–275.
18. Zaw, K., Kwik, Z. Z. S., Chang, W. Q., Islam, M. D., & Poh, T. K. (2023). A techno-commercial decision making framework for optimal district cooling system design in tropical regions. *Applied Thermal Engineering*, 220. https://doi.org/10.1016/j.applthermaleng.2022.119668
19. Brigham, E. F., & Huston, J. F. (2015). *Fundamentals of financial management* (14th ed.). Cengage Learning.
20. Boltyanshi, V., Martin, H., & Soltan, V. (1998). *Geometric methods and optimization problems*. Springer New York.
21. Yu, N., & Qiu, D. (2017). The Karush-Kuhn-Tucker optimization conditions for the fuzzy optimization problems in the quotient space of fuzzy numbers. *Complexity*, 2, 1–8.
22. https://www.spgroup.com.sg/wcm/connect/spgrp/e7707139-dd07-40c4-bdfd-ebf668691070/Q12022-
23. https://taxsummaries.pwc.com/singapore/corporate/taxes-on-corporate-income
24. Consumer Price Development in Oct 2023 (mas.gov.sg). https://www.mas.gov.sg/-/media/mas-media-library/news/consumer-price-developments/2023/inflation202310.pdf
25. https://taxsummaries.pwc.com/singapore/corporate/taxes-on-corporate-income
26. Find Space | JTC. https://www.jtc.gov.sg/find-space?type=factory%20space

4

Commercial Studies of District Cooling Systems: Financial Modeling for Investment Decisions

4.1 Introduction

Large conglomerates and corporations assess their overall financial performance by engaging a set of financial documents, including the consolidated financial statement, income statement, cash flow statement, and balance sheet. With these statements, the key financial ratios [1] are usually instrumental in the company's performance, financial stability, profitability, and growth. The analysis offers all stakeholders clear visibility into the financial performance of the company, helping in investment decisions and defining strategic plans. In the realm of engineering in modern era, particularly like the development of district cooling systems (DCS), engineers and engineering managers are in need of understanding project-based financial analysis.

This chapter discusses an in-depth explanation of the decision-making journey of DCS development. The process is segmented into four main stages before actual development and operations, as illustrated in Figure 4.1. The journey begins with the technical design. This phase demands a fusion of engineering knowledge, expertise, and experience to evaluate potential cooling demand, available resources and constraints, current deployable technologies, and environmental impacts during the preliminary analysis. It is imperative that this technical phase occurs before the financial analysis to ensure that the DCS development is sustainable and advantageous from economic, social, and environmental perspectives. The optimal engineering parameters derived from the technical design stage seamlessly transition into the financial analysis stage, forming the foundations of the project's financial model. This model evaluates key financial metrics such as internal rate of return (IRR) [1], net present value (NPV) [2], and the project's payback period [3]. Favorable financial metrics pave the way for the project proposal to be presented to senior management for their decision-making. It is anticipated that multiple iterations of engineering design and financial analysis will be

DOI: 10.1201/9781032670164-4

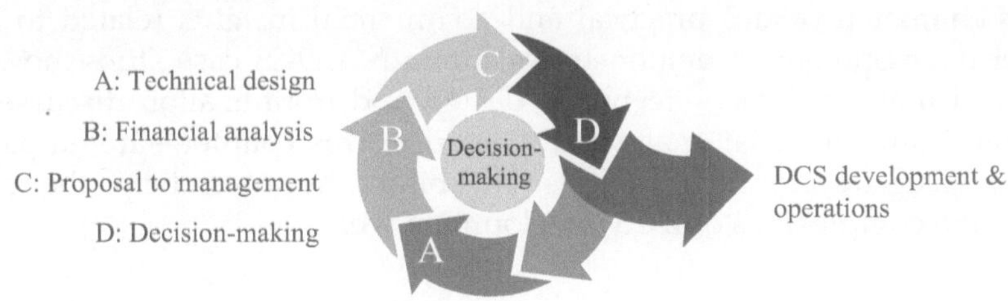

FIGURE 4.1
A typical techno-commercial decision-making framework.

conducted before finalizing the decision. This iterative process aims to optimize the engineering design and maximize financial returns. Consequently, both stages are crucial to the decision-making process. The detailed commercial and environmental benefits of the DCS system are elaborated in Chapter 6, while Chapter 3 discusses the optimal technical design procedures.

For technical design and financial analysis, the following information are important.

i. **Problem Statement Formation**: This is the fundamental stage of the project. A clear, concise, and well-defined problem statement sets the direction of the project, ensuring the objectives are in harmony with commercial goals.
ii. **Data Collection**: In this data collection stage, vital data shall be gathered. The data comprises of commercial data, technical data, environmental factors, externality, and potential operational and performance data of DCS as per the design discussed in Chapter 3. Knowing what data is needed and how to collect it is crucial, as it informs the design and implementation of the DCS.
iii. **Making Educated Assumptions**: It is often impractical to obtain all necessary information due to commercial and/or practical constraints. Educated assumptions arising from industry know-how, subject matter expertise, empirical and even theoretical knowledge always bridge the information gaps and help in determining the project behavior under different scenarios. These important expertise and knowledge help to assess inherent risks and to mitigate them.

The well-defined problem statement, data gathering, and pragmatically grounded assumptions pave the way to the financial analysis for the project. This chapter merges these steps of financial evaluation with technical parameters highlighted in Chapter 3, taking into account both initial capital expenditure (CAPEX) and long-term operating expenses (OPEX). Since DCS system design is deeply rooted in engineering, prior knowledge of DCS concept presented in Chapters 2 and 3 is vital before dealing with financial analyses.

This chapter provides practical and commercial insights related to DCS project development. It demonstrates through a DCS case study how the proposed methodologies—technical design and optimization discussed in Chapter 3 and financial analysis discussed in this chapter—are employed in actual DCS development projects. This bifocal approach helps the reader handle the complexities of DCS development well.

4.2 Brief Financial Understanding for an Engineer

In this modern era of a fast-changing and complex environment, engineering is rapidly evolving. The challenges and demands faced by engineers have ascended to unprecedented heights compared to those of previous era. Engineers must cultivate a competitive edge and embrace multi-dimensional capabilities in today's global market. Financial knowledge and commercial expertise become essential attributes to be established [4] in the engineering profession.

As an engineer involved in the financial modeling of a project, a grasp of certain financial concepts enhances the financial assessment of the project. Several key concepts are particularly related to the financial modeling of a project for decision-making. The aim of this session is to briefly discuss some financial terms and knowledge as follows:

a. **Cost of capital** represents the expenses a company incurs due to providing or borrowing funds for its operations or business activities. It could comprise two main cost components—cost of company's debt and equity. The cost of debt could be considered as the interest rate for borrowed funds, while the cost of equity refers to the expected returns of investors. **Weighted average cost of capital (WACC)** is an adjusted cost of capital of both debt and equity, factoring in their respective proportions. WACC is, in fact, the minimum rate of return that the company needs to satisfy its business activities, operations, and investors.

b. **Time value of money (TVM)**: The concept of TVM [5] is essential and foundational in financial analysis. It posits that today's available cash holds greater value than an equivalent amount of cash in the future. It is because of its ability to generate earnings from time to time. This concept is widely implemented in the calculation of project net present value (NPV) and internal rate of return (IRR), which helps in decision-making of the investment. NPV and IRR are components of financial metrics to evaluate the profitability of an investment/project.

c. **Discounted cash flow (DCF) analysis**: Comprehending DCF [6] analysis is imperative as DCF facilitates the cornerstone in the calculation of financial matrices such as TVM, IRR, NPV, and future cash flow estimation. With the integration of DCF into the financial analysis, cash flows at various points in time can be discounted to define equivalent present values. This process inherently incorporates the TVM principle. Therefore, DCF analysis facilitates more precise and accurate financial evaluations. The general formula of discounted cash flow is given as

$$PV = \frac{FC_t}{(1+i)^t} \tag{4.1}$$

where FC_t is future cash value at tth year, PV is the present value, and i stands for discounted rate.

EXAMPLE 4.1

To illustrate the discounted cash flow concept, consider a project named "Project-K" having cash flows as given in Table 4.1—Initial investment of $1,000 at present, future annual cash flow from Year 1 to Year 4, while the cost of capital is considered 10%. All the future cash flow can be converted into the present value using Eq. (4.1). Therefore, the future cash flow of Year 1 becomes $PV = \frac{\$200}{(1+10\%)^1} = \$181.81$ and that of Year 2 becomes $PV = \frac{\$3{,}000}{(1+10\%)^2} = \$247.93$; see Table 4.1 for more details.

TABLE 4.1

Present and future cash flows of Project-K

Year	*Year-0 (Present time)*	*Year-1*	*Year-2*	*Year-3*	*Year-4*
Cash flows	*-$1,000*	*$200*	*$300*	*$400*	*$500*
	$181.81	←			
	$247.93		←		
	$300.53			←	
	$341.51				←

d. **NPV**: NPV of a project is the present value of all future cash flows generated by the project. In other words, NPV demonstrates how much future earnings/cash flow is worth at present by discounting the future cash flow using a specific discount rate. The NPV can be calculated as

$$NPV = \sum_{t=0}^{n} \frac{FC_t}{(1+i)^t} \tag{4.2}$$

where FC_t is expected future net cash flow at tth year, and i is discount rate or project cost of capital. The project is decided to be commercially feasible if the NPV turns out to be positive.

EXAMPLE 4.2

Here NPV of "Project-K" is analyzed. To find the NPV of "Project-K," first, the present value of each future cash flow FC_t is calculated using Eq. (4.1) and 10% cost of capital. Second, sum all discounted cash flow using Eq. (4.2):

$$\begin{aligned} NPV &= \frac{FC_0}{(1+10\%)^0} + \frac{FC_1}{(1+10\%)^1} + \frac{FC_2}{(1+10\%)^2} + \frac{FC_3}{(1+10\%)^3} + \frac{FC_4}{(1+10\%)^4} \\ &= \frac{-\$1{,}000}{(1+10\%)^0} + \frac{\$200}{(1+10\%)^1} + \frac{\$300}{(1+10\%)^2} + \frac{\$400}{(1+10\%)^3} + \frac{\$500}{(1+10\%)^4} \\ &= -\$1{,}000 + \$181.81 + \$247.93 + \$300.53 + \$341.51 \\ &= \$71.78 \end{aligned}$$

At 10% cost of capital, the NPV of Project-K is positive and it is $71.78. Therefore, Project-K is creating commercial values.

e. **IRR**: Another important financial parameter is IRR and a way to measure the efficiency of the investment. It is a discount rate at which the NPV of all cash flows from the investment become zero:

$$\underbrace{\sum_{t=0}^{n} \frac{FC_t}{(1+IRR)^t}}_{\substack{\text{Sum of Present Value} \\ \text{(PV) inflows} \\ \text{and PV outflows}}} = 0 \tag{4.3}$$

where IRR stands for internal rate of return, that is, has to be calculated via trial-and-error technique analytically [7]. However, it can be easily programmed using some programming language and/or Microsoft Excel application. In simple explanation, IRR is an annualized effective compounded rate that shows how the investment potentially generates money in the form of profit or returns. If IRR for the investment is more than the company's WACC, the investment is enticing and adding values to the company.

EXAMPLE 4.3

Continuing Examples 4.1 and 4.2, the IRR of Project-K is now calculated using Eq. (4.3):

$$\frac{FC_0}{(1+IRR)^0}+\frac{FC_1}{(1+IRR)^1}+\frac{FC_2}{(1+IRR)^2}+\frac{FC_3}{(1+IRR)^3}+\frac{FC_4}{(1+IRR)^4}=0$$

Using trial and error, IRR 12.83% is found. This is illustrated clearly in Figure 4.2.

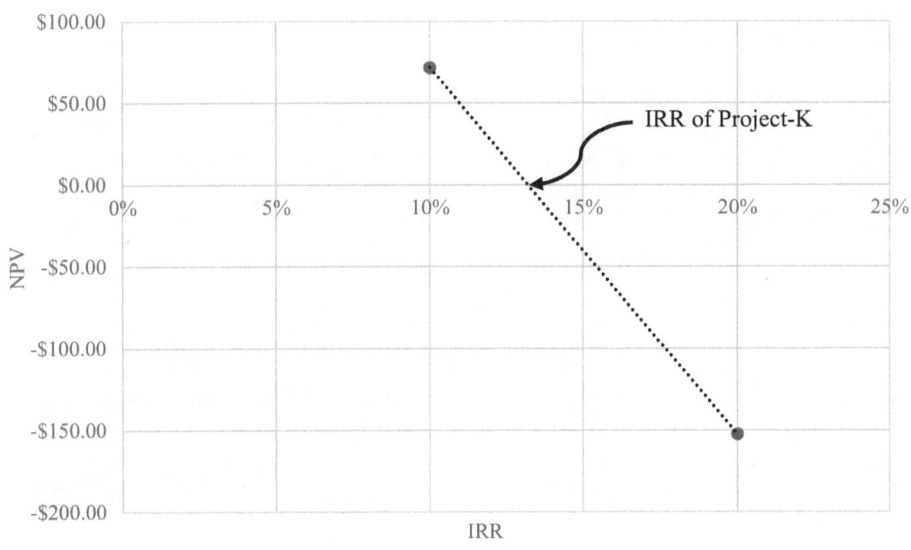

FIGURE 4.2
Calculation of IRR by trial-and-error method.

f. **Payback period** is the time duration for an investment/project to capture its initial cost, that is, the time when the cash inflows are equal to the cash outflows. Figure 4.3 shows the annual cash flow of Project-K. Figure 4.4 represents the cumulative cash flow and the payback period or breakeven, which occurs at 3.2 years.

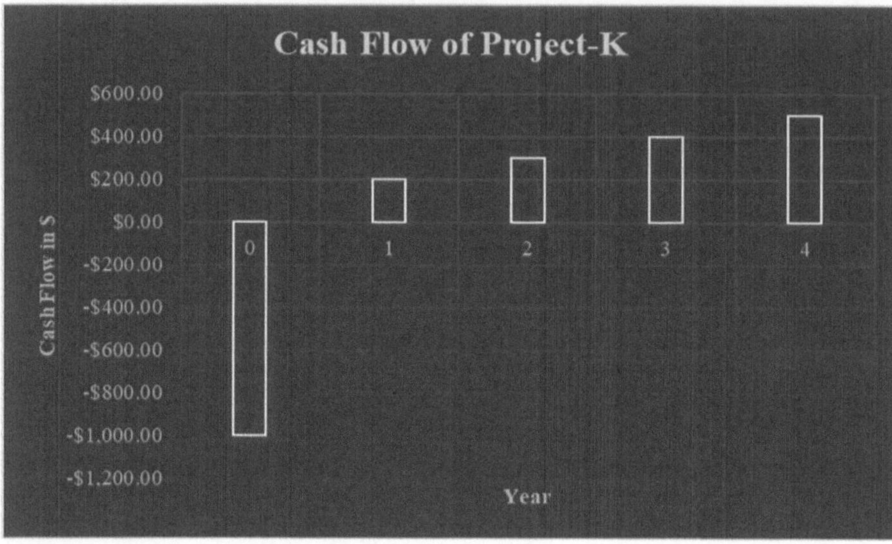

FIGURE 4.3
Annual cash flow of Project-K.

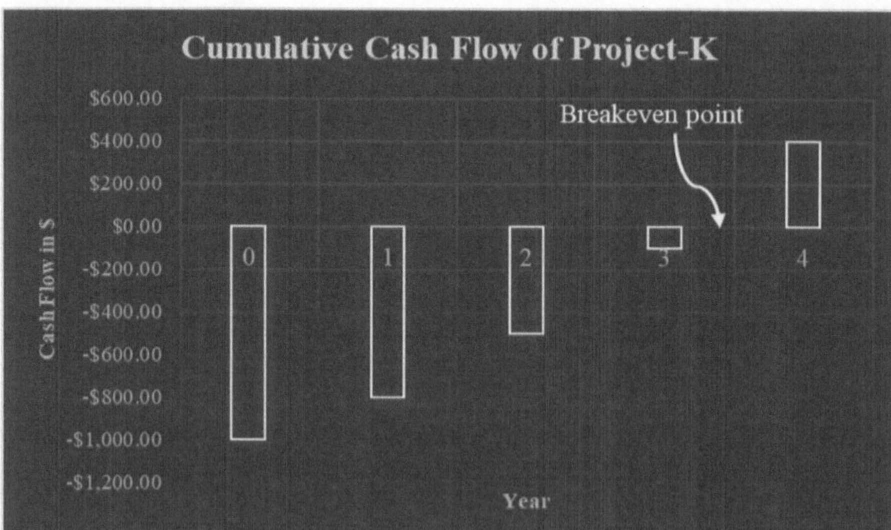

FIGURE 4.4
Cumulative cash flow of Project-K.

g. **Cash Flow Forecasting**: The ability of forecasting future cash flow is essential in financial analyses. This involves future revenue generations, anticipated demand, and operational expenses over the project's lifecycle. Industry know-how and experience play crucial roles in this informed forecasting exercise.

While this book covers foundational financial parameters, it is crucial to understand that many other financial aspects are not addressed here. Readers are suggested to delve deeper into the financial concepts through dedicated finance theory and literature [1, 8–10]. However, the theories and tools provided in this book offer fundamental and sufficient means to conduct robust financial modeling and decision-making about an investment/project as an engineer. Grasping these concepts enables us to harmonize the technical aspects of an investment/project with the desirable financial objectives.

4.3 Real-Life Setting of a DCS Project Development

In real-life situation, the development of DCS projects is capital-intensive and, therefore, typically undertaken by large corporations. The virtually indefinite existence of corporations makes them well-suited for such hefty investment projects, as the return on investment and payback period can be extended over a long duration.

To enhance comprehension of the technical and financial analyses, the case studies presented in Chapter 3 are financially analyzed here. The problem statement, data collection, and well-grounded technical assumptions are detailed in Section 3.3. This chapter concentrates on financial analyses.

4.3.1 Formulation of Project Financial Analysis

After obtaining the technical parameters, a financial model is developed to evaluate the project's commercial viability. This model defines the NPV, IRR, and the payback period of the investment. The financial model encompasses revenues, CAPEX, OPEX, depreciation of the DCS plant, and income tax components, ensuring a realistic representation of business-related scenarios [11]. The annual CAPEX is calculated using Eq. (3.10) of Chapter 3. The annual OPEX, $C_{OPEX,T}$, includes the cost of electric energy $C_{OPEX-1,t}$ as per Eq. (3.6) of Chapter 3, maintenance costs $C_{Mtn,t}$ as per Eq. (3.12), water consumption cost $C_{water,t}$, manpower/operator expenses $C_{MP,t}$, and overhead cost $C_{OH,t}$:

$$C_{OPEX,T} = C_{OPEX-1,t} + C_{Mtn,t} + C_{water,t} + C_{MP,t} + C_{OH,t}. \tag{4.4}$$

The annual cost of water is empirically defined using real operation data of a DCS plant in Singapore,

$$C_{water,t} = \kappa \times \left(TF_{water,1} \times (1+i)^{t-1} \right) \times (Q_{Total} \times 365) \tag{4.5}$$

where $\kappa = 0.0084 \, m^3/RTh$ is empirical coefficient [11], $TF_{water,1}$ is water tariff at initial year in $\$/m^3$, and Q_{Total} is daily DCS cooling demand in RTh.

The annual manpower cost $C_{MP,t}$ is given as

$$C_{MP,t} = M \times N_{people} \times \left(S \times (1+i)^{t-1} \right) \tag{4.6}$$

where M is the total number of months per year, N_{people} is the number of DCS operators, and S is the monthly salary of each staff.

The annual overhead cost $C_{OH,t}$ is determined as a percentage of the initial investment

$$C_{OH,t} = a \times C_{CAPEX} \times (1+i)^{t-1} \tag{4.7}$$

where a is a certain percentage and C_{CAPEX} is the CAPEX of DCS plant, including chiller system, thermal energy storage, and chilled water pipe network (see Eq. (3.10)). The percentage value of $a = 1\%$ is applied in this chapter. However, the interpretation of this percentage may vary according to the subjective criteria of each individual organization.

After defining the expenses, the annual revenue generated from the DCS, R_t, is expressed as a function of cooling energy and its tariff:

$$R_t = \underbrace{(Q_{Total} \times 365)}_{\substack{\text{annual cooling} \\ \text{energy sold}}} \times \underbrace{TF_{Cooling,1} \times (1+i)^{t-1}}_{\substack{\text{tariff of cooling energy} \\ \text{at year "t"}}} \tag{4.8}$$

where $TF_{Cooling,1}$ is cooling energy tariff or selling price of cooling energy in the initial year.

The operating cash flow after corporate income tax is calculated using Eqs. (4.1)–(4.8),

$$CF_t = (R_t - C_{OPEX,T} - D_t) \times (1 - i_{tax}) \tag{4.9}$$

where D_t is the straight-line depreciation $(= C_{CAPEX} / \mathfrak{N})$, i_{tax} is the corporate income tax rate in percentage, and subscript "t" ranges from 1 to $\mathfrak{N}$.

Having investment, expenses, revenue, and depreciation allows for the calculation of operating cash flow that is crucial and serves as the foundation for determining key financial metrics—NPV as in Eq. (4.10), IRR in Eq. (4.11), and the payback period.

$$NPV = CF_0 + \sum_{t=1}^{\mathfrak{N}} \frac{CF_t}{(1+k)^t} = \sum_{t=0}^{\mathfrak{N}} \frac{CF_t}{(1+k)^t} \tag{4.10}$$

where $CF_0 = -C_{CAPEX}$, and k is the WACC [1]. NPV is fundamentally a calculation that defines the monetary value of future cash flow over the economic life of an investment, that is, a DCS project in this analysis. The purpose of NPV is to assess the investment's ability to generate wealth, thereby justifying the financial attractiveness and profitability.

Another measurable profitability indicator of the project, IRR, is calculated alongside NPV, while assuming a constant reinvestment rate.

$$\sum_{t=0}^{\mathfrak{N}} \frac{CF_t}{(1+IRR)^t} = 0 \tag{4.11}$$

Finally, the payback period (together with NPV and IRR) can be determined. It is the amount of time or duration needed to recoup the initial investment. In practical scenarios, investment decision-making is typically based on the NPV and IRR. In this chapter, these parameters are utilized to validate the effectiveness of the proposed techno-commercial framework.

4.3.2 Case Studies, Results, and Discussion

The summary of the case study is given as follows, and the details can be revisited in Chapter 3.

- **Problem Statement**: A DCS system supplying cooling energy to ten buildings within a district is designed in an optimal manner.
- **Data Collection**: Technical data—cooling load demand profile for each individual building and combined cooling load profile—are collected, analyzed, and optimized. The optimal DCS design is conducted technically.
- **Making Educated Assumption**, including commercial parameters, is defined as per Table 4.2. These commercial and industrial parameters may vary depending on geographical location, resource availability, the level of economic growth, and development in each region. The parameters are given in Table 4.2. The authors have made diligent efforts to ensure that the commercial parameters are as reasonable as possible. However, they remain educated assumptions intended to support the analyses in this chapter.

TABLE 4.2

Input Parameters for DCS Design Optimization and Financial Analyses

Input Parameters		
Off-peak hours in Singapore	h_{OP}	8 hours
Peak hours in Singapore	h_P	16 hours (7:00 am – 11:00 pm period)
Peak electric energy tariff	TF_{P-1}	\$0.2491 /kWh
Contract capacity tariff [12]	TF_{CC-1}	\$13.44/kW/month
DCS system efficiency	$\eta_{DCS,1}$	0.65 kW/RT
Unit cost of thermal energy storage	U_{TES}	\$250/RTh
Unit cost of chiller system	$U_{chiller-system}$	\$3,000/RT
Unit cost for chiller maintenance	$U_{Mnt,1}$	\$50/RT/annum
Unit cost for water	$TF_{water,1}$	\$2.3/m³
Economic life of chiller system	$\mathfrak{N}$	20 years
Year-on-year efficiency degradation rate	d	0.5%
Annual escalation or discount rate [13]	i	3%[a]
Singapore corporate income tax [14]	i_{tax}	17%
Weighted average cost of capital, WACC [15]	k	7%[b]
Number of staff to operate the DCS plant	N_{people}	6
Monthly salary	S	\$3,500 per head per month
Total number of months per year (considering the bonuses for each operation staff in multiple of months)	M	15[c] (12 months for salary plus 2 months for bonus and annual wage supplement)
Rental cost [16]	$TF_{rental,1}$	\$32/m²/month[d]

Notes:

[a] Averaged Consumer Price Index (CPI) in Singapore is between 1.5% and 2.5% in 2022. The number is rounded to 3%.

[b] WACC is the weighted average cost of capital of a Singapore-based energy-related company. 7% is assumed in this book.

[c] Each operating staff receives 14 times his/her monthly salary annually with the consideration of bonuses.

[d] Rental cost could vary depending on location and types of buildings. The rental cost of \$32/m²/month is considered as an average rental cost for industrial estate.

Note: Dollar (\$) throughout this chapter refers to Singapore Dollar.

Financial Analysis of Optimal DCS Design: Using the technical design parameters from Table 3.3 and the commercial parameters listed in Table 4.2, financial metrics such as NPV, IRR, and payback period are calculated for the optimal design case, while off-peak electricity tariff is \$0.1514/kWh. The chilled water tariff charged to customers is defined in \$/RTh, and \$0.3/RTh is assumed in this book. In practice, the chilled water tariff needs to be adjusted to reflect the cost savings for customers, showing that using chilled water from DCS is beneficial. The DCS provider must have detailed discussions with each customer to understand the costs of their own cooling system, including cooling energy costs, capital expenses (CAPEX), and operational expenses (OPEX). The price offered for chilled water from the DCS should be lower and more financially advantageous for the customer. As a result, different DCS customers may have different prices for chilled water due to factors such as (i) the commercial agreements between the customers and the DCS provider, (ii) the effectiveness of the DCS marketing team, and (iii) the life cycle cost saving analysis provided to the customers. The concept of life cycle cost analysis will be explained in the Discussion section.

Figure 4.5 illustrates the annual operating cash flow generated from the optimal DCS design, based on the optimization algorithm described in Chapter 3. In Year 0, the cash flow is negative, reflecting the initial investment in the DCS plant. From Year 1 onward, the cash flow becomes positive and continues to grow until the end of the plant's economic life, demonstrating strong potential for commercial success. It is important to note that for the cash flow to remain positive in the early years, the cooling demand for the DCS must be substantial. There is a risk that the anticipated cooling demand may not be met in practice. Therefore, it is crucial to make realistic assumptions. However, in this case study, it is assumed that the anticipated cooling demand is achieved.

Figure 4.6 demonstrates the cumulative cash flow of the optimal DCS design. The breakeven point of the optimal DCS design is found after 5 years.

Analysis of Different Cases: Financial analyses for 27 cases, given in Table 3.2, are conducted to examine the optimal DCS design algorithm given in Section

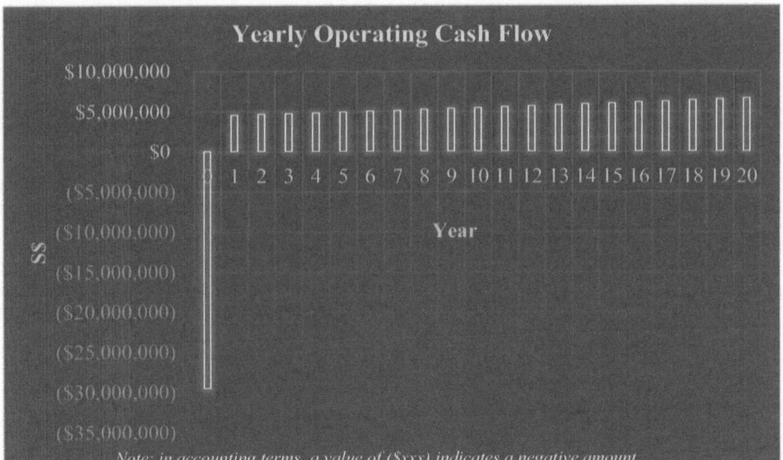

FIGURE 4.5
Operating cash flow of the optimal DCS design.

(Note: the numbers enclosed in parentheses indicate a negative value in accounting terms)

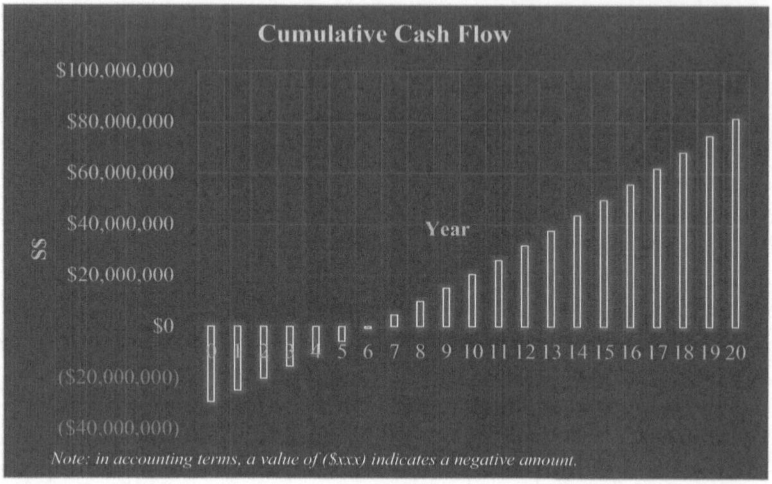

FIGURE 4.6
Cumulative cash flow of the optimal DCS design.

3.2. These 27 cases comprise three different thermal energy stage (TES) capacities, three different CAPEXs, and three variations of operational expenditures.

- The first parameter, TES capacity, includes three variations: (i) limited thermal energy storage (LTES), (ii) optimal thermal energy storage (OTES), resulting from the DCS design optimization process, and (iii) excess thermal energy storage (ETES).
- The second parameter involves investment rates of the chiller system and thermal energy storage system: the present rate, a 25% high rate and a 50% higher rate. Currently, unit cost of chiller $U_{chiller-system}$ and thermal energy storage U_{TES} are \$3,000/RT and \$250/RTh, respectively. At a 25% high investment rate, $U_{chiller-system}$ and U_{TES} become \$3,750/RT and \$313/RTh, respectively. At a 50% high investment rate, $U_{chiller-system}$ and U_{TES} become \$4,500/RT and \$375/RTh, respectively.
- The third parameter is the electricity tariff, with off-peak tariffs set at 30%, 60%, and 90% of the peak tariff. The peak and off-peak electric energy tariff is \$0.2491/kWh and \$0.07473/kWh, respectively. It implies that the off-peak tariff is approximately 30% of peak tariff. A similar calculation applies to the off-peak electric energy variations at 60% and 90% of the peak tariff.

Each parameter's three variations result in 27 unique combinations or cases.

First, nine cases are examined, encompassing three different energy storage capacities: LTES, OTES, and ETES. These cases also involve three

TABLE 4.3

Comparison of NPV, IRR, and Payback Period for Nine Cases Comprising Three Different Thermal Energy Storage Capacities and Three Different Investment Rates, Maintaining Off-Peak Electricity Tariff is 30% of Peak Electricity Tariff

$TF_{OP} = 30\% \times TF_P$ (1)	Limited TES (a)	Optimal TES (b)	Excess TES (c)
Present investment rate (A)	NPV:$29,551,772 IRR: 16.8% *Payback: 5.03 years* *(case: LTES-A-1)*	**$36,573,950** **19.3%** ***4.34 years*** ***(case: OTES-A-1)***	$33,150,144 16.8% *5.03 years* *(case: ETES-A-1)*
25% higher investment rate (B)	$21,679,919 13.0% *6.42 years* *(case: LTES-B-1)*	**$29,046,360** **15.2%** ***5.59 years*** ***(case: OTES-B-1)***	$24,435,910 13.1% *6.42 years* *(case: ETES-B-1)*
50% higher investment rate (C)	$13,808,067 10.3% 7.79 years *(case: LTES-C-1)*	**$21,518,770** **12.3%** **6.80 years** ***(case: OTES-C-1)***	$15,721,677 10.4% 7.78 years *(case: ETES-C-1)*

different CAPEX structures, while maintaining the off-peak electric energy tariff at 30% of the peak tariff. The comparison of NPV, IRR, and payback period values, provided in Table 4.3, shows that the optimal DCS design, which includes the optimal TES capacity, achieves the highest NPV, the best IRR, and the shortest payback period across various investment rate scenarios. This comparison, made with the off-peak tariff consistently set at 30% of the peak tariff, indicates that the technical optimization algorithm described in Chapter 3 effectively identifies the best design for the DCS, considering both thermal energy storage capacity and the corresponding chiller capacity.

Next, the additional nine cases are investigated, similar to the first set, but with the off-peak tariff set at 60% of the peak tariff. The NPV, IRR, and payback period values for these cases are provided in Table 4.4. Finally, the last set of nine cases, similar to the previous ones, is examined with the off-peak tariff adjusted to 90% of the peak tariff. The results for these cases are shown in Table 4.5.

Across all these investigations, the optimal DCS design consistently demonstrates the highest NPV, the best IRR, and the shortest payback period under all variations of investment rate and off-peak electricity tariff rates. These results highlight that the algorithm maintains its effectiveness even with increased investment rates, underlining the robustness and reliability of the optimization process.

In essence, the DCS optimization algorithm not only optimizes the technical design but also ensures financial viability under varying investment and operational electricity tariff conditions. This reinforces the algorithm's capability to adapt to different economic scenarios while maintaining optimal performance of the DCS both technically and financially.

TABLE 4.4

Comparison of NPV, IRR, and Payback Period for Nine Cases Comprising Three Different Thermal Energy Storage Capacities and Three Different Investment Rates, Maintaining Off-Peak Electricity Tariff Is 60% of Peak Electricity Tariff

$TF_{OP} = 60\% \times TF_P$ (2)	Limited TES (a)	Optimal TES (b)	Excess TES (c)
Present investment rate (A)	NPV: \$21,695,430 IRR:14.5% Payback: 5.8 years *(case: LTES-A-2)*	**\$26,728,485** **16.3%** **5.15 years** ***(case: OTES-A-2)***	\$21,568,689 13.7% 6.10 years *(case: ETES-A-2)*
25% higher investment rate (B)	\$13,823,577 11.0% 7.38 years *(case: LTES-B-2)*	**\$19,200,895** **12.7%** **6.58 years** ***(case: OTES-B-2)***	\$12,854,456 10.4% 7.74 years *(case: ETES-B-2)*
50% higher investment rate (C)	\$5,951,725 8.5% 8.92 years *(case: LTES-C-2)*	**\$11,673,305** **10.0%** **7.98 years** ***(case: OTES-C-2)***	\$4,140,223 8.0% 9.34 years *(case: ETES-C-2)*

TABLE 4.5

Comparison of NPV, IRR, and Payback Period for Nine Cases Comprising Three Different Thermal Energy Storage Capacities and Three Different Investment Rates, Maintaining Off-Peak Electricity Tariff Is 90% of Peak Electricity Tariff

$TF_{OP} = 90\% \times TF_P$ (3)	Limited TES (a)	Optimal TES (b)	Excess TES (c)
Present investment rate (A)	NPV: \$14,236,669 IRR: 12.1% Payback: 6.76 years *(case: LTES-A-3)*	**\$17,381,264** **13.4%** **6.20 years** ***(case: OTES-A-3)***	\$10,573,331 10.5% 7.59 years *(case: ETES-A-3)*
25% higher investment rate (B)	\$6,364,817 8.9% 8.56 years *(case: LTES-B-3)*	**\$9,853,674** **10.1%** **7.87 years** ***(case: OTES-B-3)***	\$1,859,098 7.5% 9.58 years *(case: ETES-B-3)*
50% higher investment rate (C)	\$(1,507,035) 6.6% 10.33 years *(case: LTES-C-3)*	**\$2,326,084** **7.6%** **9.51 years** ***(case: OTES-C-3)***	\$(6,855,135) 5.3% 11.53 years *(case: ETES-C-3)*

4.3.3 Life Cycle Cost Savings

Life cycle cost savings are a crucial marketing tool for persuading customers to adopt DCS. In a life cycle cost saving, it is essential to compare the overall costs for customers using DCS chilled water services versus operating their own chiller plants. Several commercial parameters, as described in Table 4.2, need to be negotiated between customers and DCS service providers until an agreement is reached. Once these parameters are adjusted, the life cycle cost savings can be calculated and compared.

For customers using DCS chilled water, the annual cost is usually the product of the chilled water tariff and total cooling energy consumption, similar to how electricity costs are calculated in many countries. In contrast, if a customer opts to operate their own chiller plant, the customer must consider cost, including water and electricity consumption, investment, staff, maintenance, and other indirect expenses. These DCS costs and self-operation costs are then compared over the entire economic lifespan or contract term to gain a clear understanding.

The DCS chilled water tariff can fluctuate based on the bargaining power of both customers and DCS service providers, with both parties aiming for a mutually beneficial agreement. It is worthwhile to note that different DCS service providers offer a different chilled water tariff structure but it is not covered in this book due to commercial sensitivity.

4.4 Conclusions

A techno-commercial decision-making framework for the DCS industry has been proposed to design an optimal DCS. The simulation and design have been successfully performed, employing the proposed objective function that optimizes technical design, investment, and operating cost. The problem statement of a DCS project is clearly defined and technical analyses have been conducted to optimize both technical and financial aspects. The project financial modeling has been developed to evaluate various parameters associated with CAPEX and OPEX. The simulation result has shown that the proposed approach is able to (i) find an optimal DCS design practically, (ii) provide the best financial return and a shorter payback period, and (iii) offer a sustainable economy due to the merit of centralized chiller plant and thermal energy storage technologies. It is found that the implementation of DCS with the proposed decision-making framework is having a promising impact on the DCS industry. However, it is important to point out that the financial analyses discussed in this chapter are project-based and intended to enhance engineers' understanding. The combination of financial analysis and technical optimization aims to help engineers assess the feasibility of DCS projects beyond the engineering perspective. Nevertheless, it should be recognized that this chapter focuses solely on financial metrics related to the project.

To summarize both Chapters 3 and 4, the following should be investigated in detail for the decision-making of DCS projects.

- The project shall start with a well-defined problem statement. This leads to the technical analyses to be performed to identify the best solutions, with the aid of engineering and subject matter expertise.
- With the support of technical analysis, financial modeling is essential to assess the annual revenue and direct and indirect cost implications of the project.
- Combining financial analysis with technical design optimization, engineers can holistically evaluate the project feasibility from both engineering and financial standpoints. Usually, financial metrics, particularly three parameters, namely, IRR, NPV, and payback period, are essential for understanding whether the project is adding values to the company's financial performance.

Overall, this approach ensures that engineers have a comprehensive understanding of both the financial and technical aspects of their projects, leading to better decision-making and project success.

References

1. Brigham, E. F., & Huston, J. F. (2015). *Fundamentals of financial management* (14th ed.). Cengage Learning.
2. Arnaboldi, M., Azzone, G., & Giorgino, M. (2015). *Performance measurement and management for engineers*, Elsevier Inc. ISBN 978-0-12-801902-3.
3. How to calculate the payback period with Excel. https://www.investopedia.com/
4. Siciliano, G. (2015). *Finance for nonfinancial managers* (2nd ed.). McGraw-Hill.
5. Goel, S. (2019). *Finance for non-finance people* (2nd ed.). Routledge.
6. Laitinen, E. K. (2019). Discounted cash flow (DCF) as a measure of startup financial success. *Theoretical Economics Letters*, 9, 2997–3020. https://doi.org/10.4236/tel.2019.98185
7. Gessinger, G. H. (2009). *Materials and innovative product development*. Elsevier Inc. https://doi.org/10.1016/C2009-0-19022-0
8. Rees, M. (2018). *Principles of financial modelling: Model design and best practices using Excel and VBA*. John Wiley & Sons, Ltd.
9. Pignataro, P. (2022). *Financial modeling and valuation: A practical guide to investment banking and private equity* (2nd ed.). John Wiley & Sons, Inc.
10. Damodaran, A. (2014). *Applied corporate finance* (4th ed.). John Wiley & Sons, Inc.

11. Zaw, K., Kwik, Z. Z. S., Chang, W. Q., Islam, M. D., & Poh, T. K. (2023). A techno-commercial decision making framework for optimal district cooling system design in tropical regions. *Applied Thermal Engineering*, 220. https://doi.org/10.1016/j.applthermaleng.2022.119668
12. Electricity tariff revision for the period 1 July to 30 Sep 2023. spgroup.com.sg
13. Consumer price development in Oct 2023. http://mas.gov.sg/
14. https://taxsummaries.pwc.com/singapore/corporate/taxes-on-corporate-income
15. Keppel Corporation Ltd (BN4) discount rate - WACC & cost of equity - Alpha spread.
16. Find Space | JTC

5

Real-Time Operational Optimization for District Cooling Systems

5.1 Introduction

District cooling systems (DCS) include chillers, pumps, thermal energy storage tanks, and cooling towers (CT), known for their high efficiency, sustainability, and cost-effectiveness. Therefore, energy-efficient DCS plant operation becomes increasingly crucial to reduce energy consumption, greenhouse gas emission, and economic cost savings.

5.1.1 Manual Operational Optimization

Typically, the optimal operation of chiller plants or cooling energy production facilities, including DCS plants, hinges on the use of energy-efficient chillers. Chillers, being the central and most energy-intensive components within such plants, play a pivotal role. Apart from selecting energy-efficient chillers, operational efficiency can be enhanced by adopting the principle of reduced compression-lift in chillers [1]. Compression-lift refers to the temperature variance between the condenser water and the chilled water as it exits the chiller unit. A higher compression-lift translates to increased energy consumption, as the chiller's compressor must elevate the refrigerant from a lower to a higher temperature and pressure. To illustrate, consider Chiller-a and Chiller-b operating at respective evaporator temperatures of $5^{\circ}C$ and $20^{\circ}C$, while maintaining a condenser water temperature of $35^{\circ}C$ at the chiller outlet, as shown in Figure 5.1. This implies that the compressor of Chiller-a, compared to Chiller-b, requires a greater compression-lift to achieve the evaporator temperature of $5^{\circ}C$ and the condenser temperature of $35^{\circ}C$, resulting in more energy-intensive operations.

Nevertheless, mechanical chillers inherently come with their own set of constraints. Enhancing operating efficiency depends on various factors, including climate conditions, chiller-lift, initial design specifications, and cooling demand. Consequently, increasing the evaporator temperature may

DOI: 10.1201/9781032670164-5

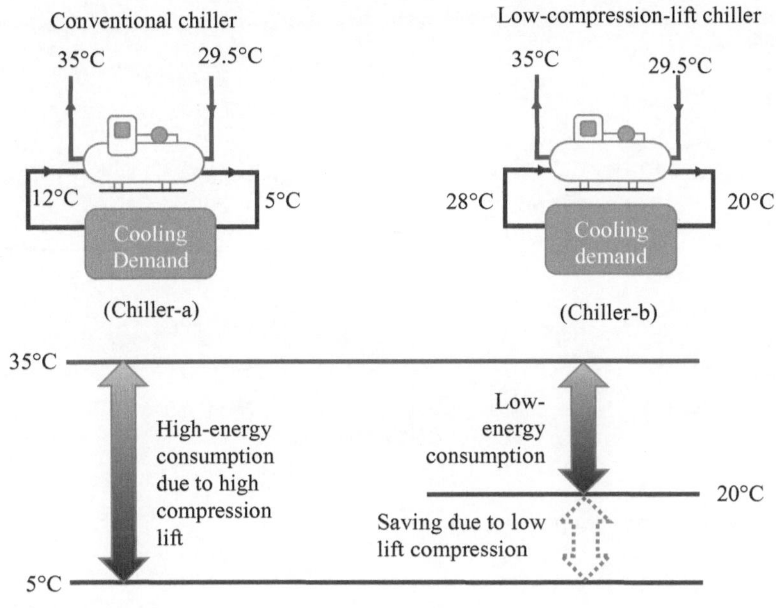

FIGURE 5.1
A comparison of high-compression-lift and low-compression-lift chillers (illustration purpose only).

not always result in improved efficiency for the chiller plant. Therefore, manual operation of a chiller plant tends to focus on optimizing chiller efficiency rather than considering the efficiencies of other plant components. It is important to note that a comprehensive understanding of typical chiller plant operations can be found in Chapter 2. Additionally, the expertise and insight of operational personnel are paramount. Furthermore, given that a chiller plant typically consists of multiple chillers and auxiliary equipment, forming what is known as a multi-chiller system, there is a growing necessity for intelligent operational optimization.

5.1.2 Research Activities and Objective

In an industry-oriented research environment, researchers concentrate on enhancing efficiency by chiller scheduling and chiller duty load management. Industry-grade chillers typically operate efficiently within a duty load range of 85%–100%. Ideally, it is observed that most mechanical chillers perform efficiently under design conditions, often referred to as full load operations, but exhibit decreased efficiency under partial load conditions, as shown in Figure 5.2. Wang et al. [2] studied the operational optimization of a chiller under different duty loads employing artificial neural network in

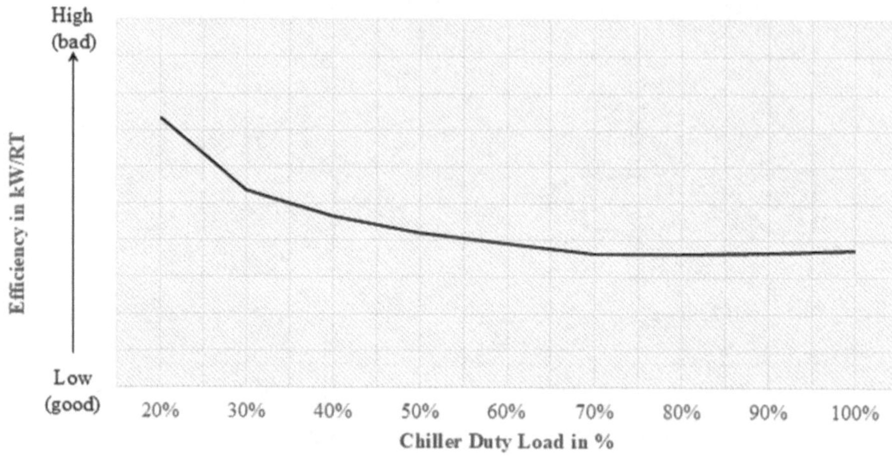

FIGURE 5.2
Chiller duty load versus its operating efficiency.

a simulation environment. Therefore, it is crucial to note that the efficiency characteristics vary among different chillers due to differences in compressors and refrigerants used. It is recommended to (i) obtain the efficiency graph from the manufacturer for greenfield/new developments and (ii) determine the chiller characteristics from the operation data for brownfield developments for specific insights. Figure 5.2 serves as a sample visual aid to solidify the understanding of chiller duty load and the corresponding efficiencies.

In practice, a chiller plant consists of multiple chiller systems, each with its own characteristics. Efficient chiller loading and scheduling are pivotal for enhancing the overall efficiency of the chiller plant. Numerous studies have investigated this aspect using a variety of optimization techniques, including but not limited to Lagrangian method and genetic algorithm [3, 4], exact algorithm [5], polynomial function, mixed integer linear programming (MILP) [6], and the exchange market technique [7, 8]. These studies demonstrate the effectiveness of optimal sequencing in improving operational energy efficiency. Aside from chiller sequencing, optimization techniques for CT have also been investigated. CT are typically modeled using heat and mass transfer equations [9] and the standard effectiveness—NTU (number of heat transfer unit) method [10, 11] to understand their characteristics. Operational optimization has been achieved by applying optimization techniques such as the particle swarm technique, revealing promising opportunities to reduce energy consumption in CT operations.

Numerous research endeavors have delved into the prediction of cooling demand [12–15], the analysis of thermal energy storage systems [16], as well as chiller sequencing and scheduling. These investigations have predominantly taken place within simulation environments, with the simulation results extensively validated against real-world operational data, showcasing reliable and promising performance.

However, the optimization approach should emphasize a holistic perspective rather than merely focusing on segmental aspects of chiller operations. Zaw et al. [17] have demonstrated the practical implementation of real-time smart operational optimization of DCS, resulting in significant energy savings. The comprehensive optimization encompasses the integration of demand prediction, chiller system, and CT optimization. The cooling demand prediction holds particular significance for real-time operational optimization of cooling plants. By leveraging predicted cooling demands, operation strategies—chiller scheduling, chiller duty load definition, pump speed modulation, and CT operations—can be defined. This facilitates energy-efficient and effective real-time management. Despite these advancements, practical implementation of these solutions has been limited. Several factors contribute to this, including the perception that manual optimization suffices for conventional chiller plants with modest energy-saving potentials, the relatively limited popularity of intelligent algorithms for operational optimization until early year 2000s, and the requirement of economies of scale. DCS, however, inherently possess economies of scale due to their district-level developments, thereby justifying investments in smart operational optimization. The content of this chapter provides a comprehensive elucidation of holistic optimization strategies for DCS. Additionally, it includes detailed insights into the implementation of these strategies in three real-world district cooling plants.

5.2 Methodology

This section concentrates on improving the daily operation of a DCS by employing mathematical models and establishing correlations through machine learning techniques. As outlined in the preceding section, the cooling demand of the DCS plant is forecasted for the next 24 hours. Based on this predicted demand, the most energy-efficient combination of chillers with their respective duty loads is determined, along with optimal pump and CT operations tailored to the prevailing weather conditions.

To accomplish this, each individual component of the DCS plant is modeled using a set of algorithms and methods that enable computers to learn from both data and physical principles. By incorporating these models into the optimization algorithm, the optimal operational strategy for the DCS plant is defined for each 30-minute interval.

5.2.1 Smart Algorithms in Brief

Certainly, there exist numerous types of machine learning and artificial intelligence techniques, and the selection of a particular technique primarily hinges on factors such as the nature of the application, the availability of data, the expertise of technical personnel, and considerations regarding scalability. Various machine learning techniques and their applications are listed in Table 5.1. The list is not exhaustive.

While numerous characterization techniques and optimization algorithms exist, the focus of this section is on neural networks, MILP, and sequential quadratic programming. These methods are chosen due to their widespread use in the cooling industry and the forthcoming discussion on practical implementations in subsequent sections.

5.2.2 Neural Network

A neural network (NN) employs a set of algorithms to forecast and identify relationships within a dataset, mirroring the functionality of neurons in the human brain. Illustrated in Figure 5.3, a standard NN model comprises layers of neurons or nodes, including an input layer, hidden layers, and an output layer, customized by the user to match specific applications. Neurons undergo training via backpropagation, while forward propagation is employed to ascertain the ultimate output of the NN.

Forward Propagation: Forward propagation denotes the process by which input data undergoes transformation to yield the final output. As illustrated in Figure 5.3, input data progresses from the input layer neurons to hidden layer neurons and subsequently to output layer neurons. Neurons within the input layer transmit the input data directly to the hidden layer without alternation. From the hidden layers onward, data undergoes modification across layers until the desired outputs are obtained. Within each neuron of the hidden and output layers, input data is subjected to a summation process, wherein interconnection weights are multiplied to produce a single value, subsequently passed through an activation function, see Figure 5.4. The summation process is represented as

$$z_j^l = \sum_{i=1}^{n} w_{ji}^l \xi_i^{l-1} + b_j^l, \tag{5.1}$$

where z_j^l is the neuron value of the layer "l" before introducing into the activation function, w_{ji}^l are the weight values (defined by the back propagation technique) from "ith" neuron of layer "l-1" to "jth" neuron of layer "l", ξ_i^{l-1} is the input value to the layer "l" or the output value from "ith" neuron of layer "l-1", and b_j^l is the bias term.

TABLE 5.1

Different Machine Learning Techniques and Optimization Techniques

Technique	Explanation and Applications
Regression technique	A statistical technique is used to analyze the relationship between one or more independent parameters. Different types of regression models are available. The technique is primarily employed for predictive tasks such as forecasting in predictive modeling [18], capturing equipment characteristics [19, 20], and evaluating risks to predict the probability of incidents [21].
Neural networks (NNs)	NNs are a form of artificial intelligence technique derived from the function of the human brain. They learn from large datasets through the training process, enabling them to predict outcomes accurately. This technique is often deployed in applications such as engineering inverse problems [22, 23], characterizing a piece of equipment, image and pattern recognition [24, 25], prediction [26], and optimization [27]. A variety of NN is available for a wide range of applications across various industries.
Genetic algorithm (GA)	GA is developed from the process of natural selection and genetics. It is a stochastic technique composed of three main probabilistic operations—selection, crossover, and mutation. In other words, it is a population-based technique that mimics the nature of evolution to find optimal outcomes. Similar to NN, a variety of GA techniques are found in different applications, including engineering [28, 29], robotic and control systems [30], and machine learning processes.
Particle swarm optimization (PSO)	PSO draws inspiration from the collective behavior of birds or fish, where individuals work together to achieve a common objective. In PSO, candidate solutions are referred to as particles. Each particle navigates through the solution space to reach an optimal solution (personal optimal) and the best solution found by the entire particle group is termed the global optimal solution. PSO is a versatile optimization technique widely utilized in various optimization problems [31]. However, in general, it is important to note that PSO does not assure convergence to the global optimum, as its effectiveness relies on the specific application and parameter configurations.
Kalman filter technique	Kalman filter is an optimal state prediction technique comprising three stages. First, it performs prediction work by estimating the current state of the system based on the previous state. Second, it carries out correction work to refine the state estimation, considering uncertainties in both the predicted state and measurements. Finally, through an iterative process, it reduces uncertainty over time. The Kalman filter is potentially used in navigation systems [32], control problems [33], and robotics. Kalman filter is generally applied for state estimations.
Mixed integer linear programming (MILP)	MILP is a popular mathematics-based optimization technique, aiming to find the optimal values of decision variables while maximizing/minimizing the objective function under a certain set of constraints. MILP is widely used across industries, contributing for processes and decision-making for energy efficiency enhancement and operational and investment cost savings.
Sequential quadratic programming (SQP)	SQP is a well-known optimization technique employed in solving different nonlinear optimization problems. Iteratively, SQP estimates the objective function while fulfilling constraints using the quadratic numerical model. SQP has often been used in engineering design, process optimization, and control.

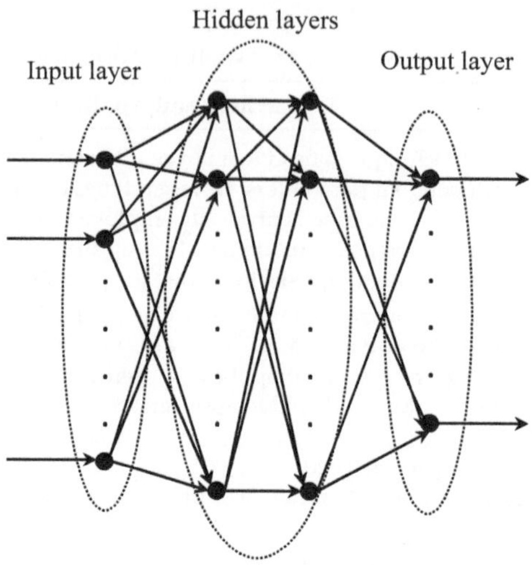

FIGURE 5.3
A typical neural network diagram with two hidden layers.

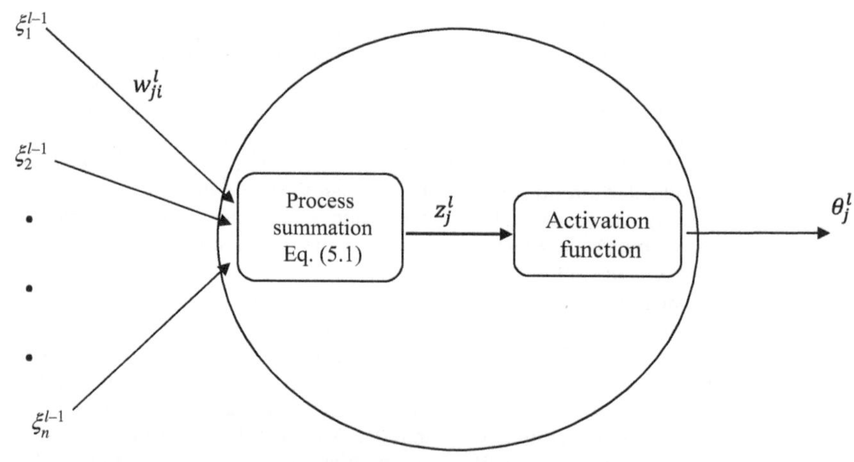

FIGURE 5.4
A typical neuron in the hidden layers and output layers.

The value z_j^l is subsequently inputted into the activation function, which adjusts the neuron's output to a comprehensible value for the next neuron in line. Although numerous activation functions exist, this study exclusively incorporates the "rectified linear unit (ReLu)" activation function [34]. ReLu transforms the z_j^l value to zero if it is zero or a negative number as shown in Eq. (5.2)

$$\theta_j^l\left(z_j^l\right) = \max\left(0, z_j^l\right). \tag{5.2}$$

This ensures that the negative values are not propagated to neurons in the subsequent layer, thereby enhancing the overall accuracy of the output value θ_j^l. It is worth mentioning that the ReLU function is applicable only when the predicted output is positive. Given that our application involves predicting cooling demand, employing the ReLU function is appropriate. ReLU's capability to produce zero values during activation contributes to sparse representation in computation, leading to simplicity and cost savings in computation.

Back Propagation: The weight function specified by Eq. (5.1) is established through the process of backpropagation. This entails refining the neuron weights according to the value of the cost function, as given in Eq. (5.3), computed after each iteration using a training dataset. Backpropagation enables the minimization of errors between the predicted and actual outputs of the model. The cost function delineates the disparity between the predicted and actual outputs:

$$C = \left(\theta^{out} - \theta^{Act}\right), \tag{5.3}$$

where θ^{out} is the predicted output of the proposed model, θ^{Act} is the actual output or actual cooling demand, and C is the cost function. In this demand prediction application, the output is the cooling demand.

Subsequently, the cost function is minimized to fine-tune the weights of each neuron. This involves adjusting the weights and biases based on the gradient of the cost function while employing the gradient descent method:

$$\frac{\partial C}{\partial w_{ji}^l} = \frac{\partial C}{\partial z_j^l} \times \frac{\partial z_j^l}{\partial w_{ji}^l}, \tag{5.4}$$

where $\frac{\partial z_j^l}{\partial w_{ji}^l} = \xi_i^{l-1}$, and thus Eq. (5.4), becomes

$$\frac{\partial C}{\partial w_{ji}^l} = \frac{\partial C}{\partial z_j^l} \times \xi_i^{l-1}. \tag{5.5}$$

Similarly, the gradient of the bias function can be defined as

$$\frac{\partial C}{\partial b_j^l} = \frac{\partial C}{\partial z_j^l} \times \frac{\partial z_j^l}{\partial b_j^l}. \tag{5.6}$$

The variation of the weights and the bias offer the minimum value of the cost function

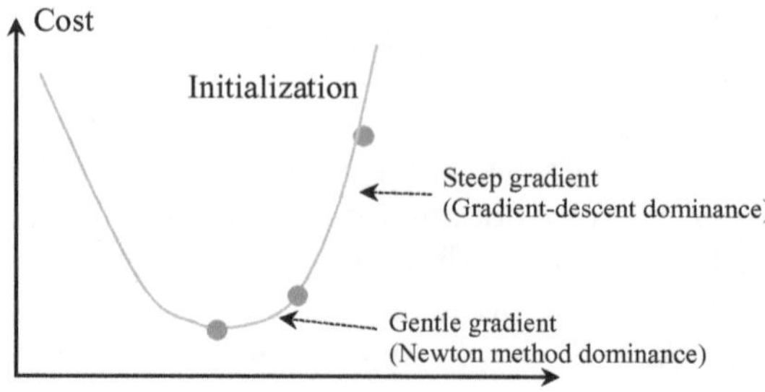

FIGURE 5.5
A simple illustration of weights and bias approximation by Levenberg-Marquardt technique.

$$w_{ji}^{l} := w_{ji}^{l} - \varepsilon \frac{\partial C}{\partial w_{ji}^{l}} \text{ and } b_{j}^{l} := b_{j}^{l} - \varepsilon \frac{\partial C}{\partial b_{j}^{l}}, \tag{5.7}$$

where ε is the learning rate for the gradient functions given in Eqs. (5.4)–(5.6). After every iteration, the gradient-decent method varies its weights and bias at the constant rate of ε. To make the process faster, the Newton method, see Eq. (5.8), is coupled with the gradient-decent method, and it is known as Levenberg-Marquardt technique. In this chapter, Levenberg-Marquardt neural network (LMNN) is utilized.

$$\left(w_{ji}^{l}\right)_{n+1} := \left(w_{ji}^{l}\right)_{n} - \left(\frac{C}{\partial C / \partial w_{ji}^{l}}\right)_{n}, \tag{5.8}$$

where the subscript "n" stands for "nth" iteration. During the backpropagation process, the gradient-decent method dominates for a steep gradient, and the Newton method dominates the gentle gradient to approach the global optimal solution as shown in Figure 5.5.

5.2.3 Mixed Integer Linear Programming (MILP)

MILP is a robust optimization method designed to handle complex problems by effectively managing numerous variables and constraints. Its versatility makes it a powerful optimization tool across various industries, including manufacturing and production, finance, logistics, process optimizations, and resource planning. By providing practical solutions to intricate challenges, MILP is widely adopted in scenarios that require precise and efficient decision-making. MILP can be formulated

$$\begin{aligned}
&\min \quad \underbrace{\mathbf{C}^T\mathbf{x}}_{\substack{objective\\function}}\\
&\text{s.t.}\\
&\mathbf{Ax} \leq \mathbf{b}\\
&\mathbf{x}\begin{cases} \in \mathbb{I}, \text{for some parameters} \\ \in \mathbb{R}, \text{for the rest} \end{cases}
\end{aligned}, \tag{5.9}$$

where **C** denotes a coefficient vector for the objective function, **x** is a decision variable vector, **A** is a coefficient matrix for the constraints and **b** is a vector of defined constraints/rules, $\mathbb{I}$ represents the space of integers, and $\mathbb{R}$ represents the space of real numbers.

In general, solving MILP problems comprises several steps: first, formulating the problem statement, then applying LP relaxation techniques to simplify the problem, and finally engaging the branch-and-cut algorithm to explore potential solutions [35]. Numerous MILP solvers [36] are available on the market to assist with these problems. For the DCS design problem, the use of MILP can be found in Section 5.3.2.

5.2.4 Sequential Quadratic Programming

Sequential quadratic programming (SQP) is an advanced optimization method customized for solving nonlinear problems. Renowned for its robustness and versatility, SQP is highly recognized in addressing constrained industrial challenges that exhibit nonlinear characteristics, akin to the capabilities of MILP. The SQP [37] problem can be formed.

$$\begin{aligned}
&\min \quad \underbrace{f_{obj}(x)}_{\substack{objective\\function}}\\
&\text{s.t.}\\
&g(x) = 0, \text{ i.e., equality constraint}\\
&h(x) \geq 0, \text{ i.e., inequality constraint}\\
&x \text{ refers to decision variable(s)}.
\end{aligned}, \tag{5.10}$$

The nonlinear problem, see Eq. (5.10), is then tackled by forming a quadratic approximate of the Lagrangian function [38]

$$L(x,\lambda,v) = f_{obj}(x) + \lambda^T g(x) + v^T h(x), \tag{5.11}$$

where λ and v refer to the vectors of Lagrangian coefficients for equality and inequality constraints, respectively. Subsequently, the quadratic subproblem is created using quadratic approximation, and the subproblem is handled iteratively as

$$\begin{aligned}
&\min \ \underbrace{\nabla f_{obj}(x_k)^T d + \frac{1}{2} d\nabla^2 L(x_k, \lambda_k, v_k) d}_{\text{objective function of subproblem}} \\
&\text{s.t.} \\
&\nabla g(x_k)^T d + g(x_k) = 0, \\
&\nabla h(x_k)^T d + h(x_k) \geq 0,
\end{aligned} \tag{5.12}$$

where x_k, λ_k, and v_k are the variable(s) and Lagrangian coefficients of the kth step. The iteration process is repeated for $k = [0,1,2,\cdots]$ until the solution of the QP problem is convergent. The SQP technique is employed in the optimization of the chiller system (Section 5.3.2) and CT operations (Section 5.3.3).

5.2.5 Problem Statement: Operational Optimization

A DCS plant, as illustrated in Figure 5.6, is a sophisticated system consisting of multiple chillers, primary chilled water pumps for circulating chilled

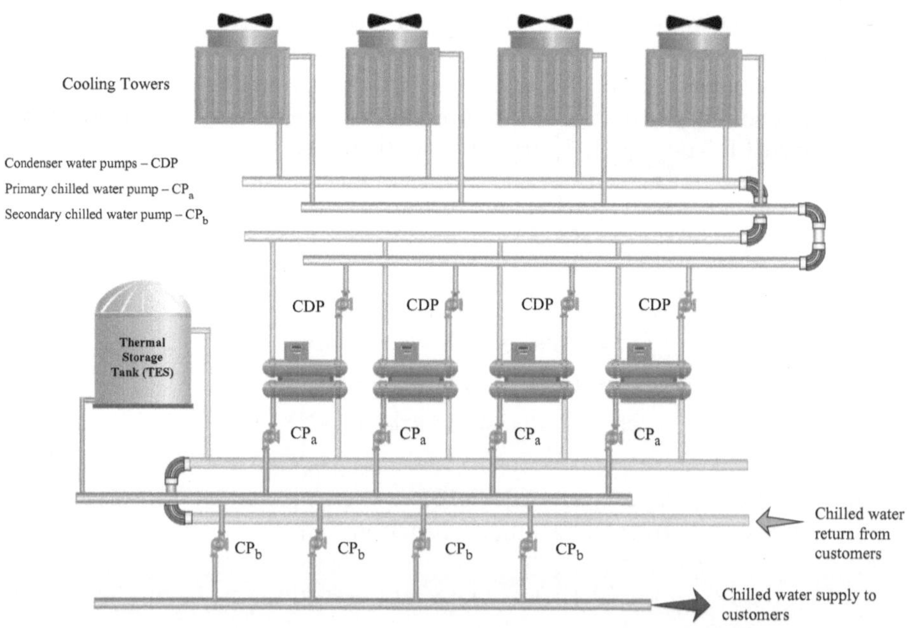

FIGURE 5.6
A typical district cooling system with four chiller systems and a thermal energy storage tank.

water from the chillers (CP_a), secondary chilled water pumps for distributing chilled water (CP_b), condenser water pumps (CDP), CT, and thermal energy storage (TES) tanks. DCS plants are renowned for their high-energy efficiency, environmental sustainability, and economic benefits. Consequently, enhancing the energy efficiency of DCS plant operations is increasingly crucial to curtail energy consumption, minimize greenhouse gas emissions, and achieve cost savings. To address this, a real-time optimization technique is discussed both theoretically and practically. This technique's effectiveness and potential for energy conservation are demonstrated.

5.2.6 Proposed Optimization

The proposed optimization is named three-layer-DCS optimization and is made up of three primary modules: (i) the cooling energy demand prediction and guidance module (PGM), (ii) the optimization module for the chiller system (CS-OPM), and (iii) the optimization module for the CT system (CT-OPM). These interconnected modules work together to create a unified and comprehensive optimization system.

Initially, the PGM module approximates the cooling energy demand for the subsequent 24 hours, which serves as the starting point for CS-OPM. Leveraging the cooling demand information, the PGM module optimizes the operation of chillers within the DCS system and devises a chiller dispatch schedule for the forthcoming 24 hours. Subsequently, CS-OPM refines the operations of chillers and pumps in real-time, while CT-OPM focuses on facilitating energy-efficient CT operations. This energy-efficient optimization aims to minimize fluctuations in energy consumption and is termed three-layer-DCS optimization, as illustrated in Figure 5.7.

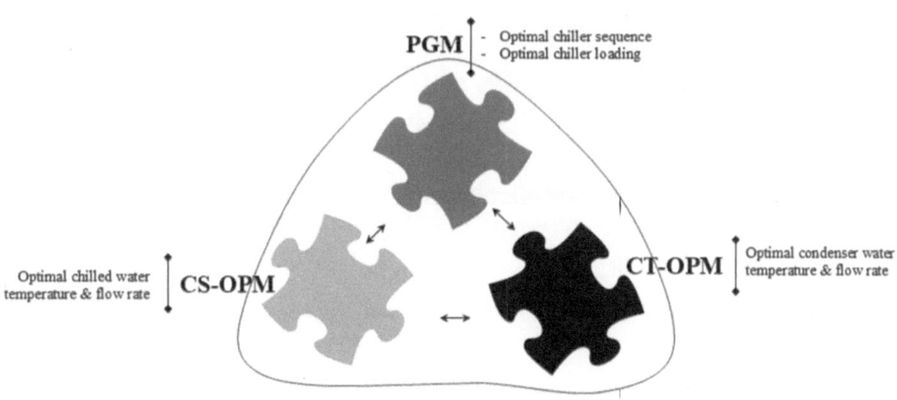

FIGURE 5.7
Three-layer-DCS optimization comprising PGM, CS-OPM, and CT-OPM.

5.2.7 Cooling Energy Demand Prediction and Guidance Module (PGM)

LMNN and data preparation with one-hot encoding: The cooling demand data for DCS customers are gathered and monitored at half-hour intervals. These data are directly influenced by the local weather conditions, including humidity, dry bulb temperature, and wet-bulb temperatures over various time periods. Correlations between these weather variables and cooling demand are analyzed for a particular year. Among these variables, only dry-bulb temperature exhibited a significant correlation with cooling demand, as it represents the ambient air temperature impacting air conditioning requirements. After establishing the relationship between cooling demand and factors such as dry-bulb temperature, day of the week, and time of day, the input parameters μ of the LMNN model are given as

$$\mu = (\xi_1, \xi_2, \cdots, \xi_{57}) \tag{5.13}$$

where ξ_1 refers to the dry-bulb temperature in °C, ξ_2 to ξ_9 correspond to the days of the week, and ξ_{10} to ξ_{57} denote the different times within a day, as shown in Figure 5.8. In this specific cooling demand prediction, one-hot encoding functions [39] to categorize the data into a vector format, preparing it for application in the proposed method. This encoding approach efficiently handles and captures the characteristics of cooling demand across different times and days, preparing the data for training the LMNN model.

Operation Guide: A well-established MILP [6] method is employed to optimize chiller selection and load distribution, based on forecasted cooling demand. The MILP model is structured to provide detailed operational

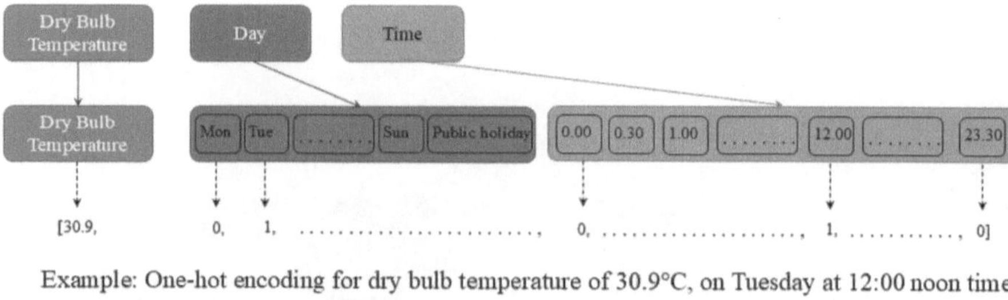

FIGURE 5.8
A sample data categorization using one-hot encoding. A data point with a dry-bulb temperature of 30.9°C recorded at 12:00 noon on Tuesday, 1 December 2015, is transformed into a one-hot encoding vector where $\xi_1 = 30.9$, $\xi_3 = 1$ (indicating Tuesday), $\xi_{34} = 1$ (indicating 12:00 noon) with all other values set to "0" [17].

guidance for the DCS over a full day, with adjustments made every 30 minutes to ensure accurate and efficient planning:

$$\min\sum_{j=1}^{48}\left(\sum_{i=1}^{n}\left(x_i\left(Q_{ec,i}\right)\times Q_{ec,i}\right)\right)_j$$

s.t.

$$\left(\left(\sum_{i=1}^{n}Q_{ec,i}\right)_j+Q_{TES,j}\right)\geq Q_{DEMAND,jhr} \tag{5.14}$$

$$Q_{TES,j}\geq 0,\text{for discharging}$$
$$Q_{TES,j}\leq 0,\text{for charging}$$
$$x_i\left(Q_{ec,i}\right)=\eta_{ec,i}$$
$$Q_{LL,i}<Q_{ec,i}<Q_{UL,i}$$

In the context of Eq (5.14), $x_i\left(Q_{ec,i}\right)=\eta_{ec,i}$ represents the efficiency of ith chiller in kilo-watt per refrigeration ton (kW/RT) within the DCS chiller group, producing $Q_{ec,i}$ in RT. The term $Q_{TES,j}$ denotes the cooling energy being charged to or discharged from the TES tank, while $Q_{LL,i}$ and $Q_{UL,i}$ signify the minimum and maximum load limits of the chillers, set according to operational demands. Chiller efficiency, measured in kW/RT, reflects the amount of electrical power utilized to generate a ton of refrigeration, which is roughly 3.517 kW. A lower kW/RT value indicates higher efficiency, as it uses less electricity for the same cooling output. The core of reducing energy consumption in chiller operations lies in solving Eq. (5.14). Constraints such as $Q_{LL,i}<Q_{ec,i}<Q_{UL,i}$ and $\left(\left(\sum_{i=1}^{n}Q_{ec,i}\right)_j+Q_{TES,j}\right)\geq Q_{DEMAND,jhr}$ must be met. By optimizing Eq. (5.14), the cooling loads are distributed efficiently across the various chillers in the DCS system. Furthermore, DCS system performance is boosted with the integration of the chiller system optimization module (CS-OPM) and the CT optimization module (CT-OPM).

5.2.8 Optimization of Chiller Systems

The CS-OPM combines the operation of chillers, primary and secondary chilled water pumps (also known as chilled water distribution pumps), and TES. It utilizes data provided by the PGM module, as illustrated in Figure 5.9, to enhance the DCS system performance.

The CS-OPM operates according to the chiller sequencing set by the PGM, making adjustments every 30 minutes to control chilled water temperature

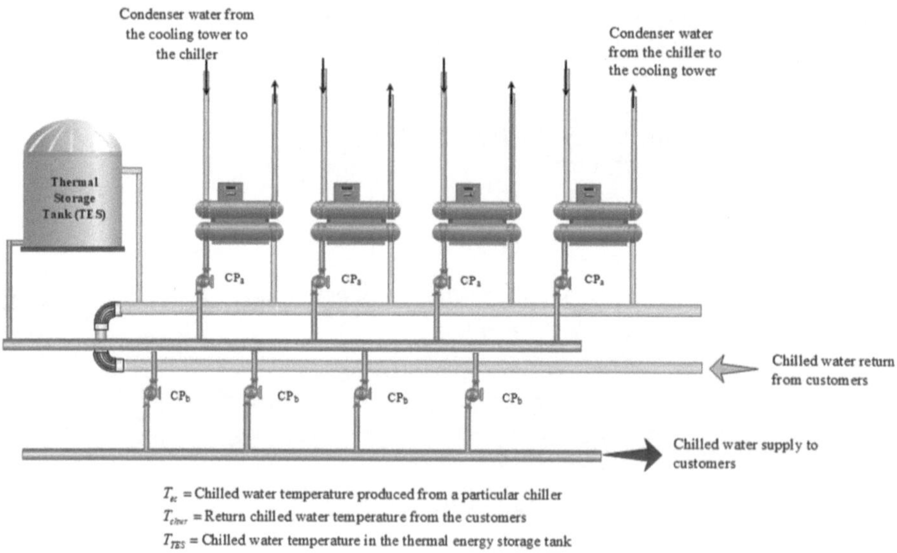

FIGURE 5.9
A typical chiller system with its chilled water and condenser water arrangement.

and flow rates. These modifications impact both chiller load and pump efficiency. The decision to use a 30-minute interval is based on operational expertise, which suggests that industrial-grade chillers typically require 15–30 minutes to adjust their output. Furthermore, considering the dynamics of DCS operations and energy management practices, the 30-minute interval is considered optimal for smooth operational performance. The objective function of CS-OPM is

$$G^* = arg\left(min\left(\Delta\mathbb{E}_{EC} + \Delta\mathbb{E}_{CPa} + \Delta\mathbb{E}_{CPb}\right)\right), \tag{5.15}$$

where $\Delta\mathbb{E}_{EC}$, $\Delta\mathbb{E}_{CPa}$, and $\Delta\mathbb{E}_{CPb}$ are the shifts in energy consumption for the driven chillers, chilled water pumps, and distribution pumps, respectively. These fluctuations are calculated based on the mathematical models that govern the operation of the chillers and pumps.

Mathematical Model of Chillers: Accurately modeling the electric energy consumption of the chillers in the DCS plant is essential because each chiller operates with distinct characteristics, which significantly impact overall energy efficiency and performance. For an *i*th chiller from the DCS chiller clusters, it can be defined as

$$E_{ec,i} = \eta_{ec,i} \times Q_{ec,i}, \tag{5.16}$$

where $E_{ec,i}, \eta_{ec,i}$ and $Q_{ec,i}$ are the electric energy consumption in kW, chiller efficiency in kW/RT, and cooling energy production in RT of *i*th chiller. The variation in electric energy consumption of *i*th chiller, $\Delta E_{ec,i}$, is determined as

$$\Delta E_{ec,i} = \left(\Delta\eta_{ec,i} \times Q_{ec,i}\right) + \left(\eta_{ec,i} \times \Delta Q_{ec,i}\right), \tag{5.17}$$

where

$$\begin{aligned} \Delta E_{ec,i} &= \left(E'_{ec,i} - E_{ec,i}\right) \\ \Delta\eta_{ec,i} &= \left(\eta'_{ec,i} - \eta_{ec,i}\right), \\ \Delta Q_{ec,i} &= \left(Q'_{ec,i} - Q_{ec,i}\right) \end{aligned} \tag{5.18}$$

The terms $\eta'_{ec,i}$, $E'_{ec,i}$, and $Q'_{ec,i}$ represent efficiency, energy consumption, and cooling demand from the previous calculation step, whereas $\eta_{ec,i}$, $E_{ec,i}$, and $Q_{ec,i}$ correspond to these values for the current step.

The chiller's efficiency at a given step, $\eta_{ec,i}$, is calculated as

$$\eta_{ec,i} = \eta_{base,i} + \left(k_{ec,i} \times \Delta T_{ec-base,i}\right), \tag{5.19}$$

where $\eta_{base,i}$ represents the reference efficiency of the chiller at the reference chilled water temperature $T_{ec-base,i}$, $k_{ec,i}$ is a coefficient derived from the operation data, and $\Delta T_{ec-base,i} = T_{ec-sp,i} - T_{ec-base,i}$ denotes the deviation of chilled water temperature set point from the reference temperature. The cooling energy, $Q_{ec,i}$, is defined as

$$Q_{ec,i} = \mathfrak{c} \times C_p \times F_{ec,i} \times \left(T_{chwr} - T_{ec,i}\right). \tag{5.20}$$

In Eq. (5.20), $\mathfrak{c}$ is a constant having $\approx \dfrac{1}{3.517} RT/kW$, C_p is the specific heat capacity in kJ/(kg °C), $F_{ec,i}$ denotes the chilled water flow rate measured in kg/s, T_{chwr} is the temperature of chilled water returning from the cooling demand point to the chiller, and $T_{ec,i}$ is chilled water temperature generated from the chiller, in °C.

Adopting the fundamental of partial derivatives, the variation in chiller operating load, $\Delta Q_{ec,i}$, is expressed as a function of variations in supply chilled water temperature $\Delta T_{ec,i}$, return chilled water temperature ΔT_{chwr}, and chilled water flow rate $\Delta F_{ec,i}$,

$$\Delta Q_{ec,i} = \mathfrak{c} \times \left[\left(C_p \times \Delta F_{ec,i} \times \left(T_{chwr} - T_{ec,i}\right)\right) + \left(C_p \times F_{ec,i} \times \left(\Delta T_{chwr} - \Delta T_{ec,i}\right)\right)\right], \tag{5.21}$$

and the variation in chiller efficiency, using Eqs. (5.18) and (5.19), is presented

$$\Delta\eta_{ec,i} = \left(\eta_{base,i} + \left(k_{ec,i} \times \underbrace{\Delta T'_{ec-base,i}}_{T'_{ec-sp,i} - T_{ec-base,i}}\right)\right) - \left(\eta_{base,i} + \left(k_{ec,i} \times \underbrace{\Delta T_{ec-base,i}}_{T_{ec-sp,i} - T_{ec-base,i}}\right)\right). \tag{5.22}$$
$$= k_{ec,i}\left(T'_{ec-sp,i} - T_{ec-sp,i}\right)$$

where $T'_{ec-sp,i}$ is the supply chilled water temperature or set-point temperature from the previous step.

By applying Eqs. (5.16)–(5.22), the variation in energy consumption with "n" chillers is established as

$$\Delta\mathbb{E}_{EC} = \sum_{i=1}^{n} \Delta E_{ec,i}, \tag{5.23}$$

while enforcing the constraints

$$\begin{aligned} F_{LL,i} &< F_{ec,i} < F_{UL,i} \\ T_{LL,i} &< T'_{ec-sp,i} < T_{UL,i} \\ Q_{LL,i} &< Q_{ec,i} < Q_{UL,i}. \end{aligned} \tag{5.24}$$

where $F_{LL,i}, T_{LL,i}$ and $Q_{LL,i}$ are lower limits and $F_{UL,i}, T_{UL,i}$ and $Q_{UL,i}$ are upper limits. These limits are determined by the operational needs and requirements of the chiller.

Mathematical Representation of Primary and Secondary Chilled Water Pumps: In the DCS system, each chiller is typically paired with its own primary chilled water pump as in Figure 5.9. The energy consumption of the pump is determined by multiplying the pump's pressure, measured in kilogram-force per square centimeter (kgf/cm^2, where 1 kgf/cm^2 equals 98.07 kPa), by the flow rate in cubic meters per second (m^3/s). By factoring in the pump's efficiency and real-time operational data, the power consumption $E_{CPa,i}$ is then calculated accordingly.

$$E_{CPa,i} = k_{CPa,i} \times \frac{H_{CPa,i} \times V_{CPa,i}}{\eta_{CPa,i}} \tag{5.25}$$

where $\eta_{CPa,i}$ and $k_{CPa,i}$ are the pump's efficiency and a conversion factor, both obtained from operation data.

The pressure of the pump, $H_{CPa,i}$, is influenced by two key parameters: the motor frequency, $Hz_{CPa,i}$, in Hz and the flow rated, $V_{CPa,i}$, in m^3/s,

$$H_{CPa,i} = a_{CPa,i} V_{CPa,i}^2 + b_{CPa,i} Hz_{CPa,i} V_{CPa,i} + c_{CPa,i} Hz_{CPa,i}^2 \tag{5.26}$$

and the pressure variations of the pump, $\Delta H_{CPa,i}$, is established

$$\Delta H_{CPa,i} = \left(2.a_{CPa,i}\Delta V_{CPa,i}V_{CPa,i}\right) + b_{CPa,i}\left(\Delta Hz_{CPa,i}.V_{CPa,i} + Hz_{CPa,i}.\Delta V_{CPa,i}\right) + \left(2.c_{CPa,i}\Delta Hz_{CPa,i}Hz_{CPa,i}\right). \tag{5.27}$$

Eq (5.27) results in

$$\Delta H_{CPa,i} = \left(2\Delta V_{CPa,i}V_{CPa,i}\right)\times \underbrace{\left[\left(a_{CPa,i}\right) + \frac{b_{CPa,i}}{2}\left(\frac{\Delta Hz_{CPa,i}}{\Delta V_{CPa,i}}\right)\frac{Hz_{CPa,i}}{V_{CPa,i}} + \left(c_{CPa,i}\frac{\Delta Hz_{CPa,i}Hz_{CPa,i}}{\Delta V_{CPa,i}V_{CPa,i}}\right)\right]}_{\zeta_{CPa,i}} \tag{5.28}$$

$$\Delta H_{CPa,i} = \zeta_{CPa,i}\left(2\Delta V_{CPa,i}V_{CPa,i}\right)$$

Engaging Eqs. (5.25)–(5.28), the variation in the electric energy consumption of the pump is expressed as

$$\Delta E_{CPa,i} = \frac{k_{CPa,i}}{\eta_{CPa,i}}\times\left(\left(\Delta H_{CPa,i}\times V_{CPa,i}\right) + \left(H_{CPa,i}\times\Delta V_{CPa,i}\right)\right), \tag{5.29}$$

and, within the DCS plant, the variation in energy consumption for "n" primary pumps is represented as

$$\Delta\mathbb{E}_{CPa} = \sum_{i=1}^{n}\Delta E_{CPa,i}, \tag{5.30}$$

Likewise, the energy usage of the secondary pumps fluctuates and the variation in energy consumption $\Delta\mathbb{E}_{CPb}$ having "m" secondary pumps is given as

$$\Delta\mathbb{E}_{CPb} = \sum_{i=1}^{m}\Delta E_{CPb,i}. \tag{5.31}$$

In the operation of the DCS, the customer demand dictates how much cooling is required at any given time, while the chillers are responsible for the generation of this cooling energy. Additionally, the TES tanks act as buffers, storing excess cooling energy when demand is low and releasing it when demand is high. To ensure smooth and efficient operation, it is essential to maintain an energy balance equation in real-time, which continuously aligns the cooling supply with demand and optimizes the use of stored energy in TES tanks.

Framework for TES Systems: TES plays a vital role in the DCS infrastructure [40]. TES systems are generally categorized into two types: sensible and latent energy storage [41]. In the DCS sector, water-based systems, which fall under sensible storage, and ice storage systems, classified as latent storage, are the most commonly utilized. Paraffin-based phase-change materials (PCM), a form of latent heat energy storage, have gained recent attention, though it is still in the early stages of development [16]. This chapter primarily focuses on the application of water-based TES. These systems are charged during off-peak hours when electricity rates are lower and discharged during peak hours to meet cooling demands. The amount of energy discharged from TES tanks is calculated on an hourly basis

$$Q_{TES,j} = c \times C_p \times F_{TES,j} \times \left(T_{chwr} - T_{TES}\right), \tag{5.32}$$

where $Q_{TES,j}$ is the TES discharged energy at jth hour, $c \approx 1/3.517\ \text{RT/kW}$, $F_{TES,j}$ is the chilled water flow rate in kg/s, C_p is the specific heat capacity in kJ/(kg °C), and $\left(T_{chwr} - T_{TES}\right)$ is the temperature difference between chilled water temperature returning from the cooling demand point and the stored chilled water temperature in the TES.

Therefore, energy balance equation at jth hour period is given as

$$Q_{DEMAND,j} = \begin{cases} \left(\sum_{i=1}^{n} Q_{ec,i}\right)_j - Q_{TES,j}, \text{for charging} \\ \left(\sum_{i=1}^{n} Q_{ec,i}\right)_j + Q_{TES,j}, \text{for discharging} \end{cases} \tag{5.33}$$

where $Q_{DEMAND,j}$ represents the cooling demand, $\left(\sum_{i=1}^{n} Q_{ec,i}\right)_j$ is the sum of produced cooling energy, "n" refers to the number of chillers, and $Q_{TES,j}$ is the charging/discharging of TES, at jth hour. Volumetric water balance equation is given as

$$V_{DEMAND,j} = \begin{cases} \left(\sum_{i=1}^{n} V_{ec,i}\right)_j - V_{TES,j}, \text{for charging} \\ \left(\sum_{i=1}^{n} V_{ec,i}\right)_j + V_{TES,j}, \text{for discharging.} \end{cases} \tag{5.34}$$

Influence of Chilled Water Supply Temperature from the DCS Plant: In a DCS plant, the chilled water temperature is usually maintained at a fixed level due to factors like the original system and chiller design, contractual

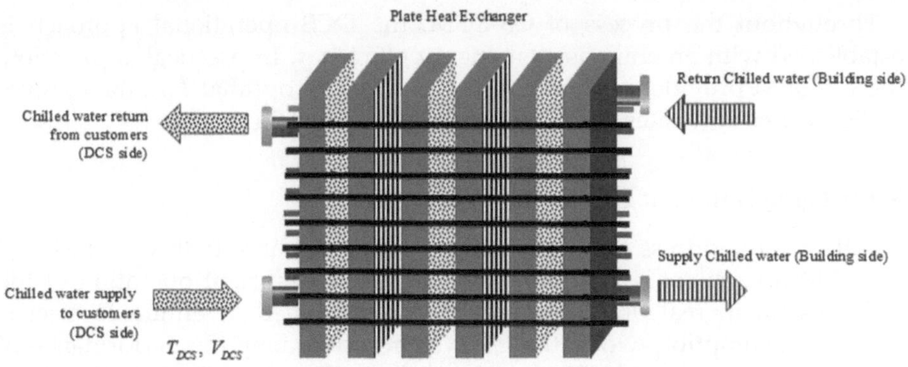

FIGURE 5.10
Energy transfer heat exchanger from the DCS to the customer buildings.

obligations with customers, and inefficiencies from manual operations. However, adjusting the chilled water supply temperature slightly presents an opportunity for energy savings. To facilitate this, a mathematical model has been created for the energy transfer station, as shown in Figure 5.10. The energy transfer station, typically a plate heat exchanger, acts as a barrier between the DCS chilled water system and the customer's building, enabling the cooling energy to be exchanged solely through heat transfer.

The term $Q_{DEMAND,jhr}$ is as a function of $T_{DCS} = T_{\ TES} = T_{ec,i}$, and the supply chilled water flow rates to the energy transfer stations are denoted as V_{DCS}

$$Q_{DEMAND,jhr} = a_{c1}T_{DCS} + a_{c2}V_{DCS} + a_{c3} \tag{5.35}$$

where the constants a_{c1}, a_{c2} and a_{c3} are derived from the historical dataset. Variations of $Q_{DEMAND,jhr}$ are then calculated and optimized to determine the corresponding changes in volumetric flow rate

$$\begin{aligned} a_{c1}\Delta T_{DCS} + a_{c2}\Delta V_{DCS} &= 0 \\ \Delta V_{DCS} &= -\frac{a_{c1}}{a_{c2}}\Delta T_{DCS}. \end{aligned} \tag{5.36}$$

The impact of the chilled water supply temperature on the return temperature is evaluated as

$$\Delta T_{chwr} = \Delta T_{DCS} - \left(\frac{T_{chwr} - T_{DCS}}{V_{DCS}}\right) \times \Delta V_{DCS}. \tag{5.37}$$

Alternations in volumetric flow in Eq. (5.36), along with chilled water temperatures in Eq. (5.37), directly influence the operation of both pumps and chillers as demonstrated in Eqs. (5.21), (5.22), and (5.29).

Throughout the process of CS-OPM, the DCS operational approach is established with an emphasis on energy efficiency. In practical application, the CS-OPM provides several outputs, such as an optimal $T_{ec,i}$, the optimal chilled water flow rates for individual chiller, and TES.

5.2.9 Optimization of Cooling Tower System

To effectively address the anticipated cooling demand, both CS-OPM and CT-OPM are activated together, ensuring energy-efficient operation of the DCS system in real time. The CT-OPM is designed to minimize electric energy consumption within the CT system, optimizing the performance of condenser water pumps, CT fans, and chillers. Therefore, the objective function of the CT-OPM module is given as a function of variations of electric energy consumption of CT fans $\Delta\mathbb{E}_{CT}$, condenser water pumps $\Delta\mathbb{E}_{CDP}$, and chillers $\Delta\mathbb{E}_{CEC}$:

$$\mathbb{F}^* = \arg\left(\min\left(\Delta\mathbb{E}_{CT} + \Delta\mathbb{E}_{CDP} + \Delta\mathbb{E}_{CEC}\right)\right), \tag{5.38}$$

while condenser water temperature, $T_{CW,ECin}$, is considered as a variable influenced by the local climate conditions, see Figure 5.11.

Temperature Threshold for Cooling Tower Operations: In the DCS plant, the energy consumption of the ith CT fan is dictated by several key factors. The fan's speed is dynamically adjusted to maintain the desired cooling water temperature $T_{CW,ECin,i}$, which is typically kept above the wet-bulb temperature to account for the CT's heat transfer efficiency. As a result, fluctuations in cooling water temperature, $\Delta T_{CW,ECin,i}$, are influenced by these adjustments:

$$\begin{aligned}\Delta T_{CW,ECin,i} = &\left(K_{11,i} \times \Delta T_{CW,ECout,i}\right) + \left(K_{12,i} \times \Delta F_{CW,i}\right)\\ &+ \left(K_{13,i} \times \Delta T_{WB}\right) + \left(K_{14,i} \times \Delta F_{AIR,i}\right),\end{aligned} \tag{5.39}$$

where $\Delta T_{WB}, \Delta T_{CW,ECout,i}, \Delta F_{CW,i}$ and $\Delta F_{AIR,i}$ are variations of wet-bulb temperature, cooling water temperature leaving from the chiller, condenser water flow rate, and fan air flow rate, respectively, while maintaining the limits of $F_{CW,min,i} < F_{CW,i} < F_{CW,max,i}$. The coefficients $K_{11,i}, K_{12,i}, K_{13,i}$ and $K_{14,i}$ are generated from the operational dataset.

Variations in Energy Consumption of CT Fans, Condenser Water Pumps, and Chillers: To enhance the efficiency of CT operations and optimize $\Delta T_{CW,ECin,i}$, the speed of CT fan is modified, and the resulting change in energy usage (measured in kW) is estimated based on the principle of the affinity law [43, 44] and fluid dynamics [42]. The affinity law establishes how airflow, electric energy consumption, and motor speed influence each other, serving as a guide to predict how changes in fan speed impact energy utilization

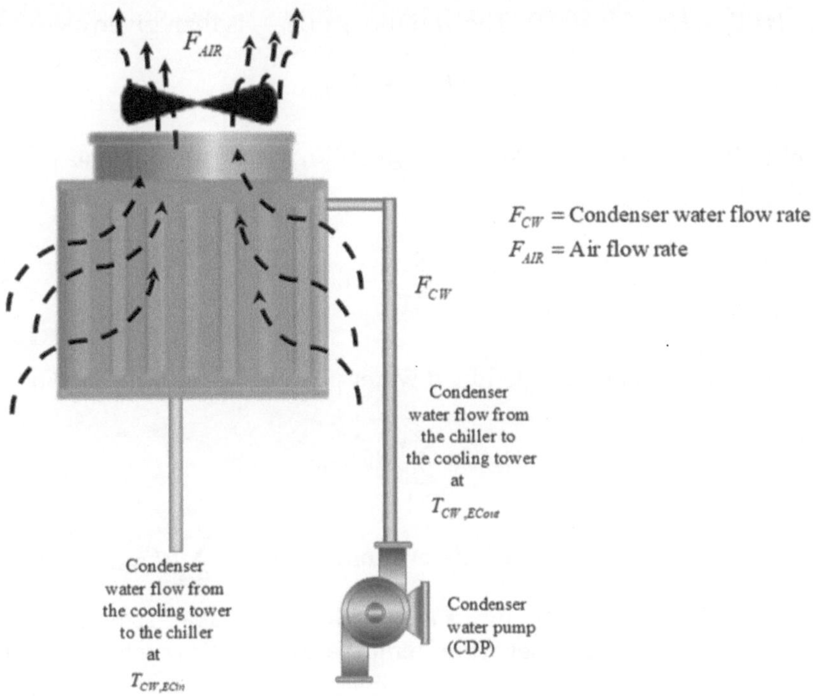

FIGURE 5.11
A typical cooling tower with a condenser water pump (the condenser water conveys the heat generated by the chillers to the cooling tower, where it is dissipated into the atmosphere through a combination of latent and sensible heat transfer within the air stream).

$$F_{AIR,i} \,\alpha\, N \text{ and } E_{CT,i} \,\alpha\, N^3 \tag{5.40}$$

where N represents the CT fan's speed which is directly related to the frequency $Hz_{CT,i}$ in Hz, and Eq. (5.40) is demonstrated as

$$F_{AIR,i} = K_{fan-1} \times Hz_{CT,i} \text{ and } E_{CT,i} = K_{fan-2} \times Hz_{CT,i}^3. \tag{5.41}$$

In Eq. (5.41), K_{fan-1} and K_{fan-2} are constants and are sourced from operating data of CT fans or the manufacturer's fan curve. This equation allows for the calculation of the fan's power consumption, $E_{CT,i}$, based on the air flow rate, using these constants:

$$Hz_{CT,i} = \frac{F_{AIR,i}}{K_{fan-1}} \text{ and } E_{CT,i} = \left(\frac{K_{fan-2} \times Hz_{CT,i}^3}{K_{fan-1}}\right) F_{AIR,i} = K_{fan,i} F_{AIR,i} \tag{5.42}$$

where $Hz_{CT,i}^{min} < Hz_{CT,i} < Hz_{CT,i}^{max}$. The variations of $E_{CT,i}$ is then given as

$$\Delta E_{CT,i} = K_{fan,i} \times \Delta F_{AIR,i}. \tag{5.43}$$

Finally, the variation in total energy utilization of CT fans, $\Delta \mathbb{E}_{CT}$, is described as

$$\Delta \mathbb{E}_{CT} = \sum_{i=1}^{n} \Delta E_{CT,i} \tag{5.44}$$

Following the operations of chilled water pumps, $\Delta \mathbb{E}_{CDP}$ is defined as

$$\Delta \mathbb{E}_{CDP} = \sum_{i=1}^{n} \Delta E_{CDP,i}. \tag{5.45}$$

In order to compute Eq. (5.38), determining $\Delta \mathbb{E}_{CEC} = \sum_{i=1}^{n} \Delta E_{ECE,i}$ is essential. The variation in electric energy usage of ith chiller, $\Delta E_{CEC,i}$, in response to changes in both the condenser water temperature and flow rate, is given as

$$\Delta E_{CEC,i} = \left(a_{CEC,i} \times \Delta F_{CW,i}\right) + \left(b_{CEC,i} \times \Delta T_{CW,ECin,i}\right), \tag{5.46}$$

where $a_{CEC,i}$ and $b_{CEC,i}$ are empirical constants. The CT-OPM is performed, utilizing Eqs. (5.44)–(5.46), and determines the optimal conditions of CT operations. Due to the complexity and nonlinearity of both CS-OPM and CT-OPM, SEQ is utilized to handle the objective functions defined in Eqs. (5.15) and (5.38).

In summary, the three-layer optimization process of the DCS is performed to ensure energy-efficient DCS plant operations, following the framework outlined in Eqs. (5.15)–(5.38). The optimization process is systematically represented in the flowchart in Figure 5.12, highlighting the key steps involved.

i. Historical data is initially gathered to build an NN model capable of predicting customer demand and to derive empirical constants for the intelligent mathematical models governing chillers, pumps, and CT.
ii. Using the PGM module, an optimal chiller dispatch schedule is formulated, focusing on enhancing operational efficiency.
iii. The predicted demand and empirical equations are integrated into the CS-OPM and CT-OPM modules, which work together to establish a real-time strategy for energy-efficient operations. This involves

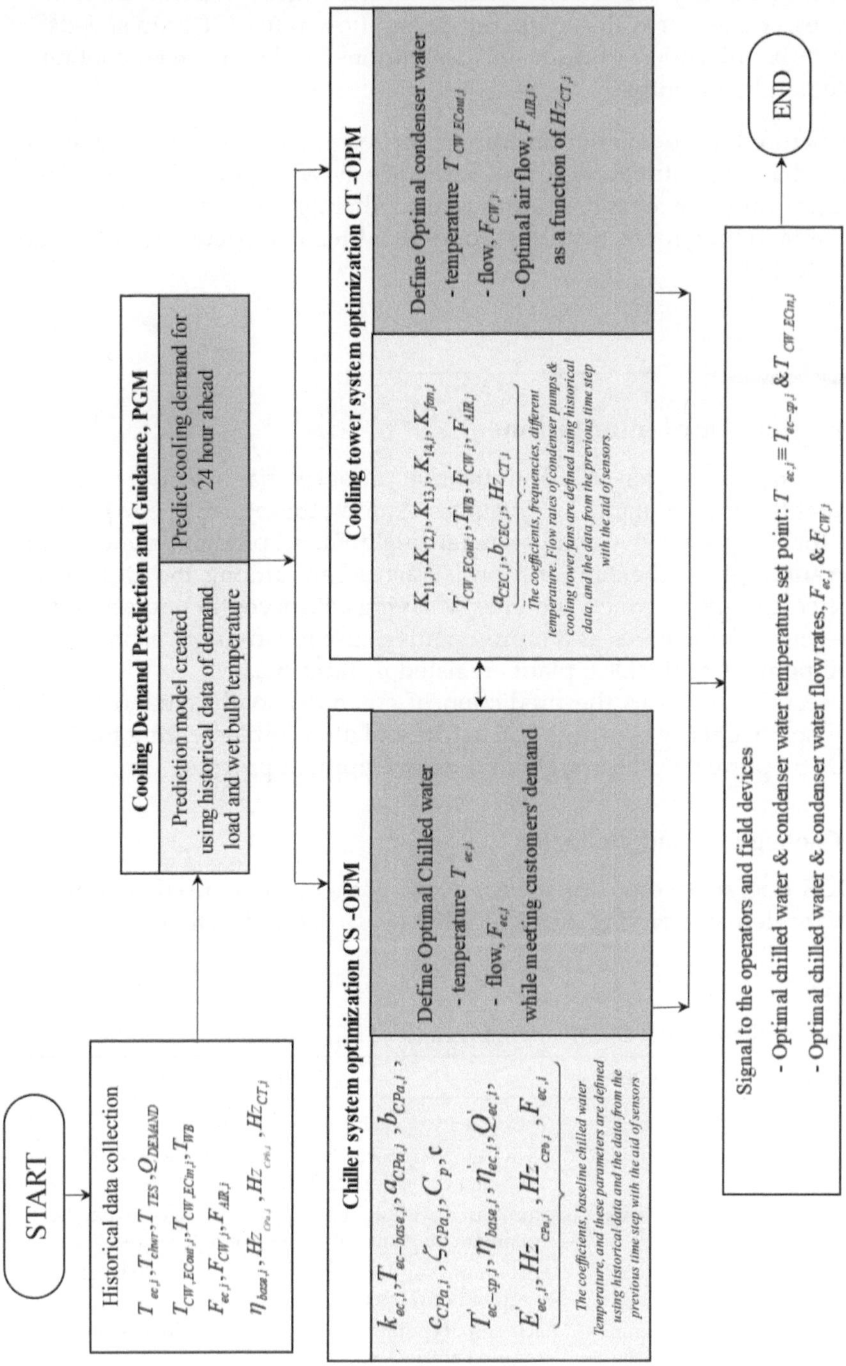

FIGURE 5.12
Three-layer-DCS optimization flowchart [17].

optimizing key parameters such as chilled water temperature, flow rates, condenser water temperature and flow rates, CT's fan speeds, and the allocation of duties across various chillers to meet cooling demand efficiently.

This structured three-layer optimization approach ensures a systematic process for achieving optimal energy performance in DCS plants. The three-layer DCS optimization process has been successfully implemented in commercial DCS plants in Singapore, with the operational results discussed in the subsequent section.

5.3 Practical Implementations

The real-time three-layer-DCS optimization proposed has been put into practice across three DCS plants in Singapore. Among these plants, one has been chosen for comprehensive analysis regarding demand prediction and energy conservation, given the similarity in infrastructure among the DCS plants and the consistent accuracy and energy savings observed in operation data. This selected plant serves as a representative case for in-depth examination. The components of the DCS plant are listed in Table 5.2.

This section discusses the prediction of customer load demand and the comparison of energy consumption before and after implementing the three-layer-DCS optimization throughout the development process.

5.3.1 Cooling Demand Prediction

The DCS operation data for a year were gathered and used to train an LMNN for demand prediction. The LMNN is structured with an input layer

TABLE 5.2

Components of an Actual DCS Plant in Singapore [17]

Components	Units	Remarks
Chillers	5 units	The capacity of each chiller is 2,000 RT.
Primary chilled water pumps	5 units	Each pump supports its respective chiller to meet the required chilled water flow.
Chilled water distribution pumps	6 units	6 units of distribution pumps are arranged in parallel to maintain the required DCS system pressure
Condenser water pumps	5 units	Each pump supports its respective chiller to meet the required flow for the cooling tower
Cooling towers	5 units	Each unit of cooling tower has 2–4 fans to meet the required air flow rate

comprising 57 neurons, followed by two hidden layers, each containing 60 neurons, and finally, an output layer consisting of a single neuron. Using actual cooling load data over a period of 2014, a parameter set P_N is established systematically using one-hot encoding. The sample parameter set P_N is defined:

$$P_N = \{\mu^1, \cdots, \mu^N\}, \tag{5.47}$$

where P_N covers every 30-minute period of weekdays and holidays, and N is dependent on half-an-hourly cooling demand data. For instance, if the dataset of a year is utilized for the training of the NN, the N value is 17,520 $\left(= \underbrace{365}_{\substack{\text{days in} \\ \text{a year}}} \times \underbrace{48}_{\substack{\text{total 30-minute} \\ \text{period in a day}}}\right)$. Eighty percent of the sample parameter set is utilized for the training of the LMNN model and the remaining 20% is used for validation of the trained LMNN. Performance of the trained LMNN model is evaluated, and the root mean square error (RMSE) [45, 46] is calculated to be 5.96, indicating an average difference of 5.96 RT between the predicted and actual cooling demand. Given that the cooling demand ranges from 1300 to 10,600 RT, this deviation is considered negligible. Figure 5.13 illustrates the output of the trained LMNN model compared with the target values for the training set, validation set, and the full sample set. With the exception of a few outliers, the model demonstrates impressive accuracy.

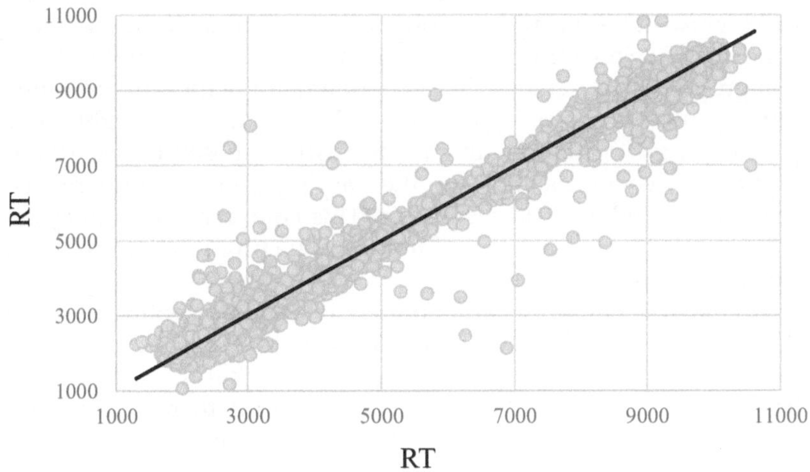

FIGURE 5.13
Linear coefficient of correlation between actual and predicted cooling demand in 2015 [17].

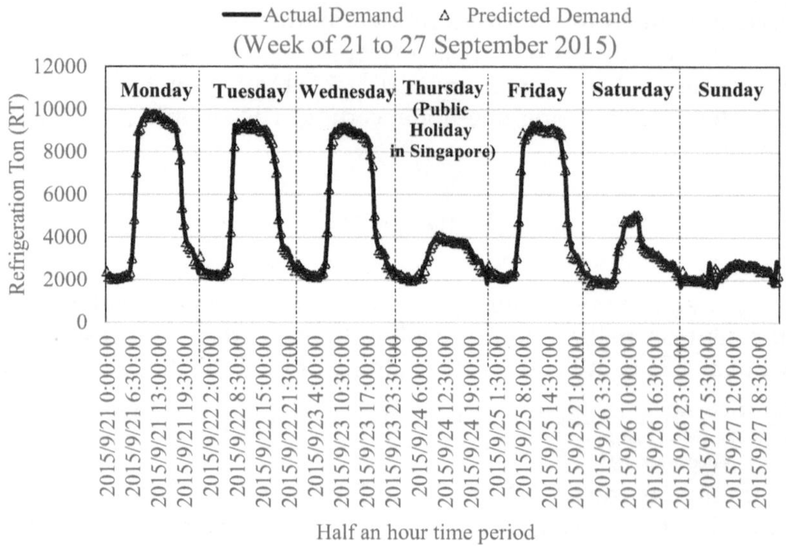

FIGURE 5.14
DCS plant's predicted demand load and actual customer demand for a week (21 to 27 September 2015) [17].

To further validate the accuracy and robustness of the cooling demand prediction model, Figure 5.14 provides a detailed comparison between the predicted and actual cooling demand over a full week, spanning from September 21 to September 27, 2015. The evaluation highlights the model's predictive performance across various days, including typical weekdays, a public holiday in Singapore, and weekends. The model's ability to consistently align predicted values with actual demand across different types of days underscores its reliability. Moreover, these results shed light on important patterns and fluctuations in cooling demand based on the day of the week, offering deeper insights into the varying cooling requirements during workdays, holidays, and weekends. This comprehensive analysis reinforces the model's practical applicability in managing cooling systems under diverse operational conditions.

A particularly notable observation is the higher cooling demand on Mondays compared to other weekdays. This can be explained by the reduced cooling activity over the weekend, which allows heat to accumulate in the building envelopes due to solar radiation. As a result, there is a spike in cooling energy demand on Mondays as the system compensates for this heat built-up, a pattern clearly illustrated in Figures 5.14 and 5.15. These insights into the week's cooling demand cycles further strengthen the model's utility in predicting and managing energy requirements.

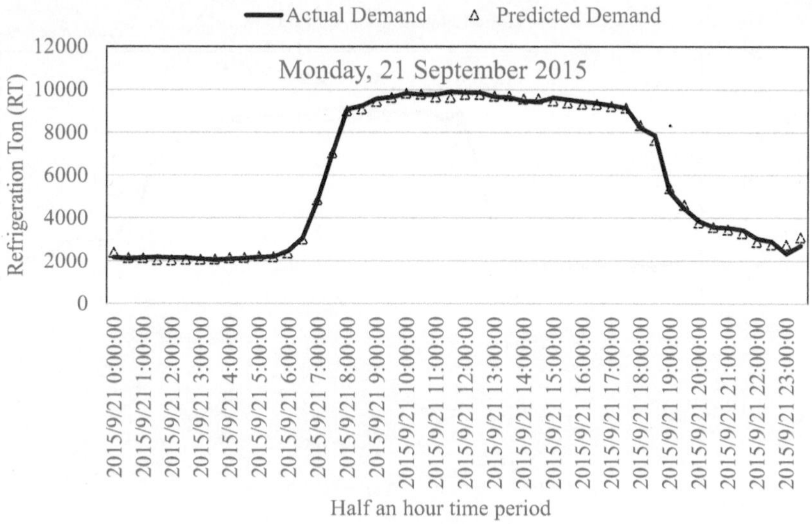

FIGURE 5.15
Actual versus predicted demand on Monday, 21 September 2015 [17].

On weekdays, it is evident that cooling energy demand tends to peak during specific hours, while significantly decreasing during off-peak periods. This trend is illustrated in Figures 5.15, 5.16, 5.17, and 5.19, clearly. In Singapore, peak hours are defined as the time between 7:00 hr and 23:00 hr, when energy consumption is highest due to the increased activity and higher occupancy in buildings. Off-peak hours, on the other hand, represent the remaining time of the day, during which energy demand is lower as a result of decreased occupancy and less activity. This distinction between peak and off-peak hours highlights the importance of strategically optimizing energy usage throughout the day to improve efficiency and reduce operational costs. Additionally, Figure 5.18 represents the cooling demand pattern observed on a public holiday—specifically, 24 September 2015, in Singapore. Cooling demand on this public holiday was notably lower compared to typical weekdays. This reduction is attributed to important factors such as reduced commercial and industrial activities, lower building occupancy, and altered energy usage patterns. Public holidays, therefore, lead to a noticeable shift in cooling energy consumption, reflecting the changes in business and occupancy dynamics. This reinforces the idea that external factors, such as holidays, can significantly influence district cooling demand patterns. The cooling demand pattern on Saturdays, illustrated in Figure 5.20, further reflects a unique trend. Between 7:00 hr and 12:00 hr, cooling demand on Saturdays typically ranges between 4,000 and 6,000 RT, significantly higher than at other times of the day. This suggests that Saturdays are often treated as partial workdays for many businesses

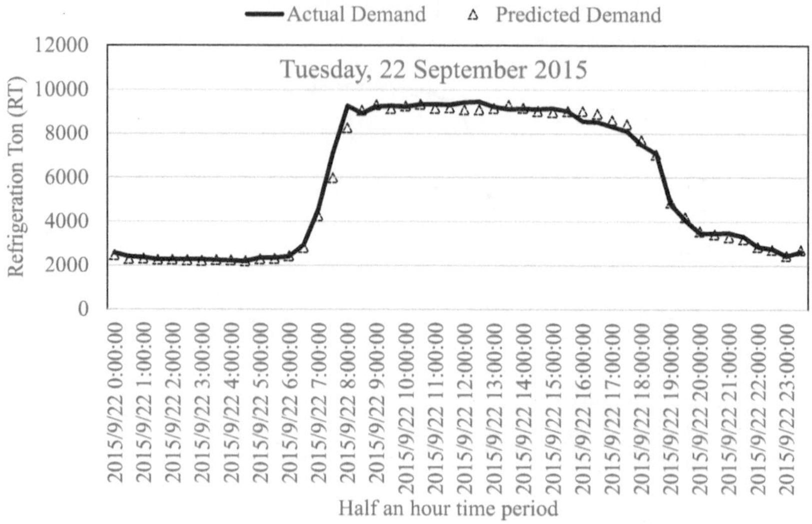

FIGURE 5.16
Actual versus predicted demand on Tuesday, 22 September 2015 [17].

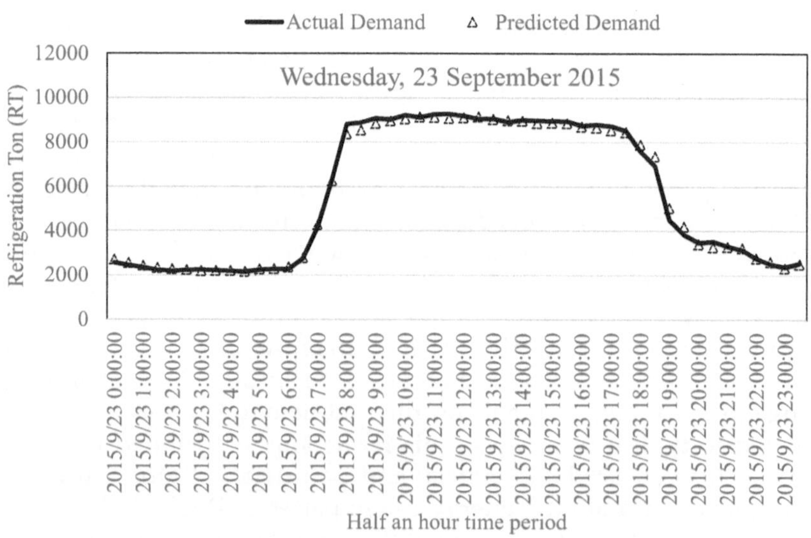

FIGURE 5.17
Actual versus predicted demand on Wednesday, 23 September 2015 [17].

in Singapore. In contrast, Figure 5.21 shows the cooling demand pattern on Sundays, which differs from that of any other day of the week. Most office buildings within the district operate at minimal capacity on Sundays, with limited functions aside from essential services such as cooling for

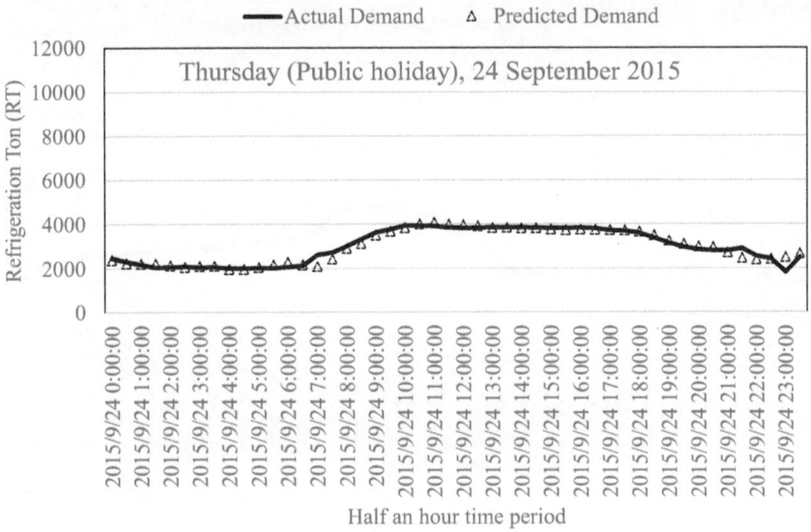

FIGURE 5.18
Actual versus predicted demand on Thursday, 24 September 2015 [17].

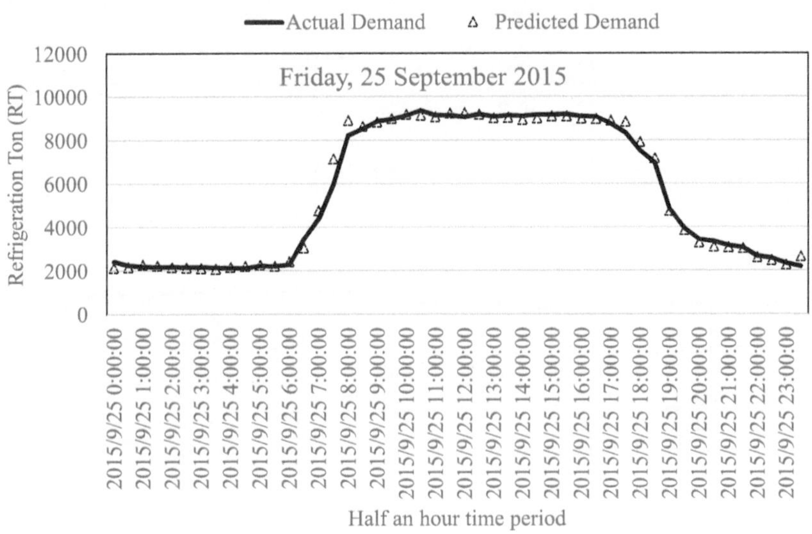

FIGURE 5.19
Actual versus predicted demand on Friday, 25 September 2015 [17].

laboratories and data server racks. As a result, cooling demand remains almost constant during off-peak hours, with only a slight increase during peak hours due to heat gain from sunlight.

These observed patterns underscore the varying cooling energy demands driven by different factors, including business activity levels, external

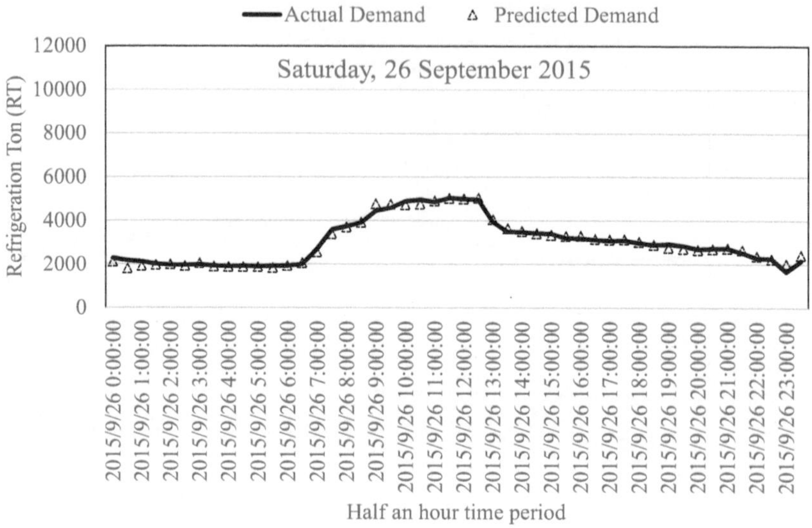

FIGURE 5.20
Actual versus predicted demand on Saturday, 26 September 2015 [17].

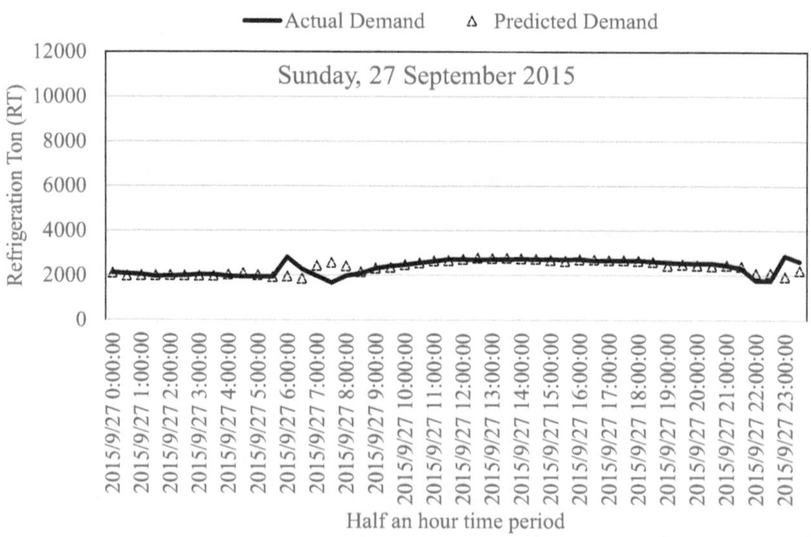

FIGURE 5.21
Actual versus predicted demand on Sunday, 27 September 2015 [17].

conditions, and occupancy rates. Understanding these fluctuations is essential for optimizing DCS to meet dynamic needs efficiently. The trained LMNN model effectively captures these diverse DCS cooling load patterns. The predicted cooling demand closely aligns with the actual demand, indicating the robustness and accuracy of the model in capturing the dynamic nature of cooling demand within the DCS zone.

While the proposed prediction technique demonstrates effective prediction capabilities, ensuring its continued accuracy necessitates the utilization of recent data for optimal performance. It is imperative to update and retrain the LMNN model at least once a year, employing a comprehensive dataset, including the entire annual load profile of the DCS as well as the relevant weather data. This approach ensures that the model remains aligned with evolving patterns and trends in cooling demand, thus enhancing its effectiveness and reliability in applications.

5.3.2 Performance of the Three-Layer Optimization

Building upon the findings discussed earlier, the output generated by PGM serves as the initialization for defining an optimal real-time operation strategy for the DCS plant, updated on a 30-minute basis. The analysis encompasses a detailed assessment of efficiency improvements before and after the implementation of optimization measures. By applying the three-layer-DCS optimization framework at the selected DCS plant, a thorough investigation of individual components and their respective efficiencies is made possible.

To safeguard proprietary information, the efficiencies of the DCS plant are normalized by dividing them with the maximum annual efficiency value, ensuring the confidentiality of sensitive operational data. The training of the PGM and the derivation of empirical coefficients required for the CS-OPM and CT-OPM are based on operational data of the year 2014.

The three-layer-DCS optimization framework was implemented in the year 2015, facilitating a direct comparison of efficiency metrics before and after implementation. The 2014 data serves as a baseline for evaluating the effectiveness of the optimization measures implemented in 2015. This comparative analysis enables a comprehensive understanding of the tangible benefits derived from the adoption of the three-layer-DCS optimization techniques within the DCS plant.

Enhancing Efficiency of Chiller Component: Figure 5.22 demonstrates the monthly efficiencies of the DCS plant. In 2014, the chilled water temperature ranged between 6°C and 6.5°C, while the condenser water temperature varied from 28.8°C to 30.5°C. It is important to note that increasing the chilled water temperature can theoretically enhance chiller efficiency by improving the chiller-lift, assuming the condenser water temperature remains constant. However, chiller efficiency is not determined solely by chilled water temperature; it is also significantly influenced by condenser water temperature. Both factors must be considered together to accurately assess chiller performance. Figure 5.23 compares the chiller efficiency for 2014 and 2015, with a significant improvement noted in 2015. This suggests that, in 2014, manual operational optimization efforts were mainly directed

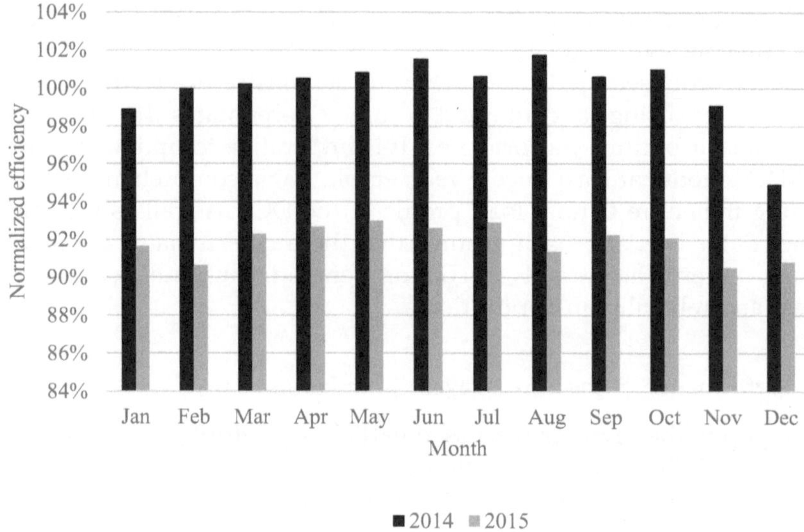

FIGURE 5.22
Monthly DCS efficiencies (2014 vs. 2015) [17].

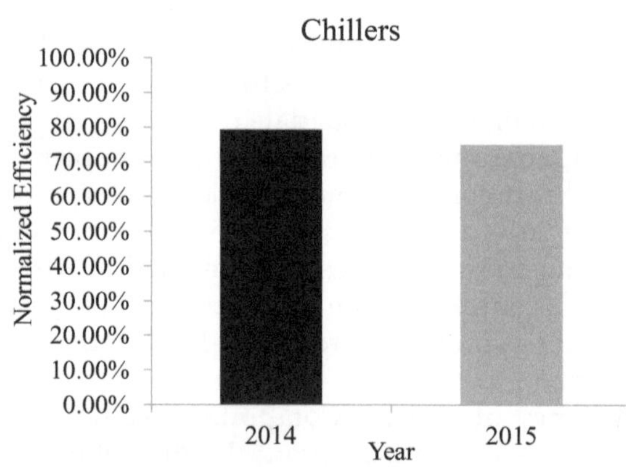

FIGURE 5.23
Overall chiller efficiency improvement for the years 2014 and 2015 [17].

at optimizing chilled water temperature without taking into full account the role of condenser water temperature. In contrast, in 2015, chilled water temperature were reduced to a range of 5°C to 6°C, leading to improved chiller component efficiency. This boost in efficiency is the result of optimizing both chilled water and condenser water temperature, enabled by the CS-OPM and CT-OPM modules. Overall, the 2015 data reveals a notable 5% improvement in efficiency compared to 2014, underscoring the positive impact of comprehensive optimization on the performance of the DCS plant.

Enhancing Efficiency of Auxiliary Components: Each of the four different DCS components—the primary chilled water pump, the condenser water pump, the CT fan, and the chilled water distribution pump—play a vital role in the system's overall performance and energy consumption. Analyzing the efficiency of each component individually provides valuable insights into how these components contribute to the DCS plant's total efficiency and operational effectiveness. Figures 5.24 through 5.27 showcase the efficiency levels of these four DCS components, highlighting their individual impact.

Figure 5.24 highlights the improvement in the performance of the primary chilled water pumps in 2015, compared to 2014. This improvement can be

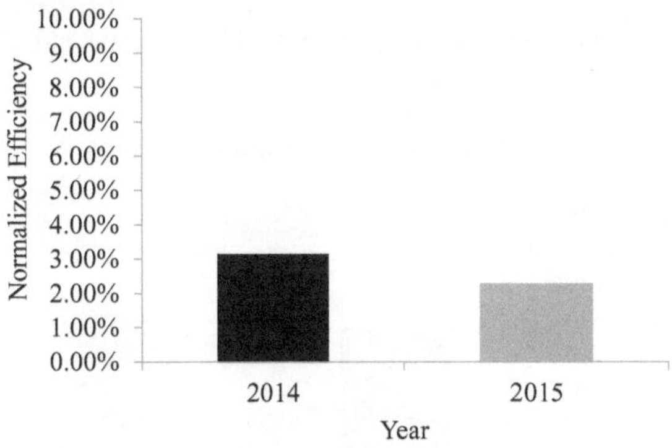

FIGURE 5.24
Overall efficiency of primary chilled water pump component for the years 2014 and 2015 [17].

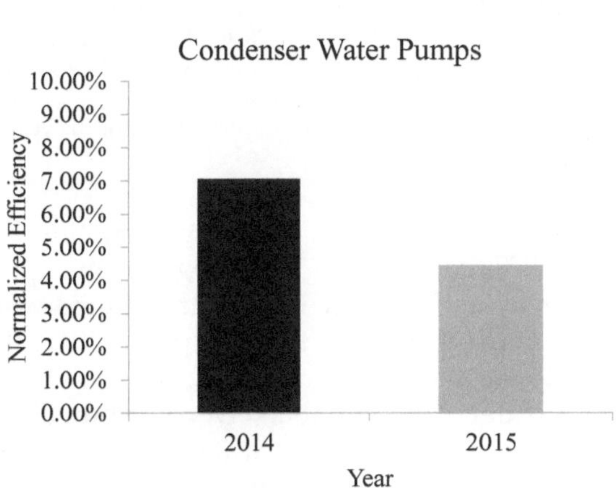

FIGURE 5.25
Overall efficiency of condenser water pump component for the years 2014 and 2015 [17].

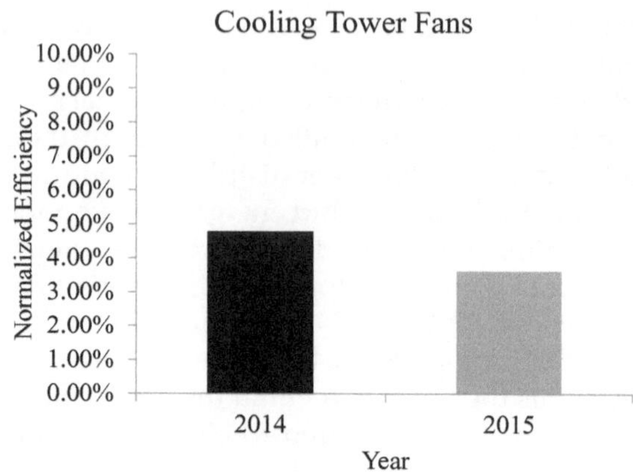

FIGURE 5.26
Overall efficiency of cooling tower fan component for the years 2014 and 2015 [17].

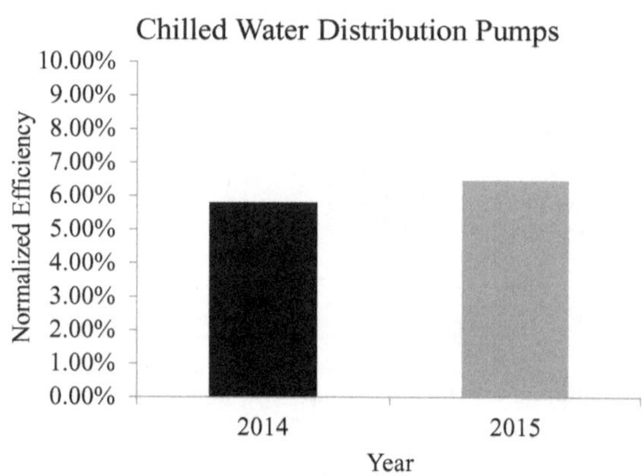

FIGURE 5.27
Overall chilled water distribution pump efficiency improvement for the years 2014 and 2015 [17].

attributed to two main factors. First, for a given cooling load, an increase in the temperature difference $\left(T_{chwr} - T_{ec,i}\right)$ as given in Eq. (5.20), results in a reduction in a lower required flow rate for the chilled water, leading to a reduction in energy consumption of the pumps. In 2015, the chilled water was produced at a lower temperature, paving the way to a larger value of $\left(T_{chwr} - T_{ec,i}\right)$. This greater differential likely contributed to the improved energy efficiency of the primary chilled water pumps. Second, implementing the CS-OPM system provided continuous optimization by regulating the chilled water flow through variable speed drives. Although these changes contributed to a better performance of the primary chilled water pump system, the overall efficiency improvement was modest, achieving an increase of 0.85%.

The performance of the CT optimization module (CT-OPM) has been evaluated by plotting the efficiencies of the CT system and condenser water pumps as shown in Figures 5.25 and 5.26. The data clearly highlights the impact of the CT-OPM on system efficiency, with both CT and condenser water pumps demonstrating a strong dependence on weather conditions. Cooler climatic conditions naturally boost the performance of CT, and by leveraging this knowledge, the CT-OPM was able to make operational adjustments that optimized condenser water temperature and flow rates. This resulted in energy efficiency improvements in both the CT fan and condenser water pump components in 2015. Figure 5.27 presents the efficiency of chilled water distribution pump or secondary chilled water pump, which shows slightly better performance in 2014 compared to 2015, with a marginal difference of about 0.68%. Given the interconnected nature of the primary and secondary chilled water pumps, the combined efficiency gains following the implementation of optimization measures have led to positive energy savings. This outlines the comprehensive nature of the optimization approach, which focuses on multiple elements of the DCS system rather than targeting individual components in isolation. Further illustrating the holistic impact of this strategy, Figure 5.28 demonstrates an overall improvement in DCS system efficiency by approximately 10% in 2015 compared to 2014. This enhancement has translated into substantial electric energy saving, estimated at 2.4 million kWh, based on an annual cooling energy production of 40 million RTh.

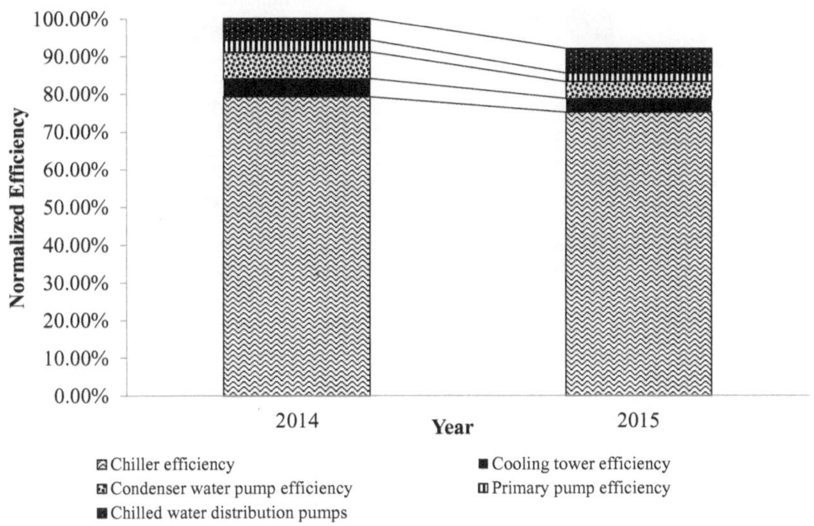

FIGURE 5.28
Overall DCS efficiency improvement of Plant 1 [17].

Overall, this comprehensive analysis reveals the significant role the proposed optimization measures have played in enhancing the energy efficiency and operational performance of the entire DCS system.

5.3.3 Implementation in Other DCS Plants

Furthermore, the application of the proposed optimization strategy has been extended to two additional DCS plants in Singapore, named Plant 2 and Plant 3, to distinguish them from the previously discussed Plant 1. Plant 2 is equipped with 12 chiller systems, each comprising associated chilled water pumps, CT, condenser water pumps, chilled water distribution pumps, and TES tanks. Similar to Plant 1, each chiller in Plant 2 has a capacity of 2,000 RT. On the other hand, Plant 3 consists of 4 chiller systems, again each with a capacity of 2,000 RT, and features TES tanks.

While detailed information about these plants is withheld due to commercial sensitivities, it is noted that the proposed three-layer-DCS optimization has been successfully implemented in both Plant 2 and Plant 3. This optimization effort has resulted in energy efficiency improvements ranging from 5% to 10%, as illustrated in Figures 5.28–5.30. These efficiency gains translate into significant energy savings, as summarized in Table 5.3.

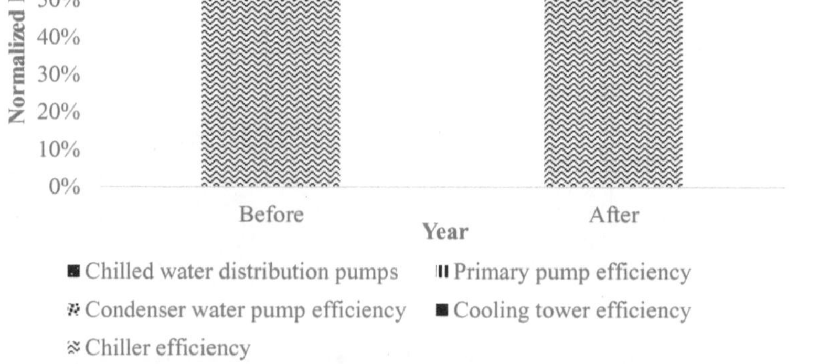

FIGURE 5.29
Overall DCS efficiency improvement of Plant 2.

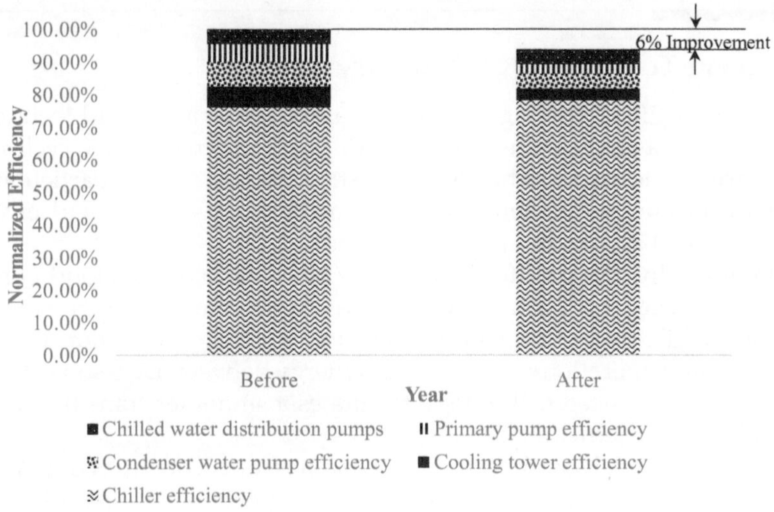

FIGURE 5.30
Overall DCS efficiency improvement of Plant 3.

TABLE 5.3
Energy Savings of Three DCS Plants

DCS Plant	Annual Electric Energy Savings
Plant 1	2.4 million kWh
Plant 2	3.7 million kWh
Plant 3	1.7 million kWh

The extent of energy efficiency improvement is influenced by the baseline efficiency of each plant, i.e. the efficiency before implementation of smart optimization. The baseline efficiency is fundamentally defined by the condition of equipment within DCS and the depth of understanding and expertise among operational personnel. The three DCS plants, in this book, possess well-maintained equipment with knowledgeable and responsible operation team, and thus, the baseline efficiency is considered quite favorable. Despite the highly efficient baseline efficiency, the implementation of intelligent optimization techniques has facilitated substantial improvements in energy efficiency. Taking into account the economics of scales, these efficiency improvements result in an annual electricity savings of approximately 7.8 million kWh. This significant impact of the optimization initiatives across the three DCS plants contributes to assessing the economic viability of investments related to the optimization.

5.4 Present Trend for DCS Industry

In the era of artificial intelligence (AI), market forces and rapid technological advancements have paved the way for the deployment of AI techniques across various industries. The DCS industry is no exception, particularly in its pursuit of energy efficiency improvements, cost savings, and environmental sustainability compared to previous era.

Technologically, the proliferation of AI/ML technologies, cloud computing, and enhanced connectivity enables us to not only optimize computer resources efficiently but also implement intelligent optimization technologies in the daily operations of DCS. This synergy between DCS subject matter expertise and advanced technology facilitates a smoother transition toward environmental sustainability.

From a commercial standpoint, DCS benefits from economies of scale, creating opportunities to invest in smart solutions that enhance operational efficiency. As a result, the integration of artificial intelligence (AI) techniques in the DCS industry drives economic growth as well as contributes to a greener and more sustainable cooling solution.

References

1. https://www.buildings.com/building-controls/article/10185904/increase-chiller-efficiency-by-reducing-lift
2. Wang, L., Lee, E.W.M., & Yuen, R.K.K. (2018). A practical approach to chiller plants' optimization. *Energy & Buildings*, 169, 332–343.
3. Chang, Y. C. (2008). Genetic algorithm based optimal chiller loading for energy conversation. *Applied Thermal Engineering*, 16, 2800–2815.
4. Chang, Y. C., Lin, J. K., & Chuang, M. H. (2005). Optimal chiller loading by genetic algorithm for reducing energy consumption. *Energy Build.*, 37 (2), 147–155.
5. Acerbi, F., Rampazzo, M., & G. De Nicolao (2020). An exact algorithm for the optimal chiller loading problem and its application to the optimal chiller sequencing problem. *Energies*. https://doi.org/10.3390/en13236372
6. Kantor, I., Robineau, J. L., Butun, H., & Marechal, F. (2020). A mixed-integer linear programming formulation for optimizing multi-scale material and energy integration. *Front. Energy Res.*, 8, 49. https://doi.org/10.3389/fenrg.2020.00049
7. Jafari, A, Khalili, T., Babaei, E., & Bidram, A. (2019). A hybrid optimization technique using exchange market and genetic algorithms. *IEEE Access*, 8, 2417–2427.
8. Sohrabi, F., Nazari-Heris, M., Mohammadi-Ivatloo, B., & Asadi, S. (2018). Optimal chiller loading for saving energy by exchange market algorithm. *Energy Build*. https://doi.org/10.1016/j.enbuild.2018.03.077

9. Ma, K., Liu, M., & Zhang, J. (2021). Online optimization method of cooling water system based on the heat transfer model for cooling tower. *Energy*, 231, 120896. https://doi.org/10.1016/j.energy.2021.120896
10. Liao, J., Xie, X., Nemer, H., Claridge, D.E., Culp, C.H. (2019). A simplified methodology to optimize the cooling tower approach temperature control schedule in a cooling system. *Energy Conversion and Management*, 199, 111950. https://doi.org/10.1016/j.enconman.2019.111950
11. Braun, J.E. (1988). *Methodologies for the design and control of central cooling plants*. http://digital.library.wisc.edu/1793/46694
12. Bird, D. N., de Wit, R., Schwaiger, H. P., Andre, K., Beermann, M., & M. Žuvela-Aloise (2019). Estimating the daily peak and annual total electricity demand for cooling in Vienna, Austria by 2050. *Urban Clim.*. https://doi.org/10.1016/j.uclim.2019.100452
13. Alhayek, M., Wadi, A., Pont, U., & Mahdavi, A. (2019). Prediction of buildings' cooling energy demand: A comparison of simulation-based and prescriptive approaches. *IOP Conference Series: Materials Science and Engineering*, 072016. https://doi.org/10.1088/1757-899X/609/7/072016
14. Vaghefi, A., Jafari, M.A., Emmanuel Bisse, Y. L., & Brouwer, B. J. (2014). Modeling and forecasting of cooling and electricity load demand. *Applied Energy*, 136, 186–195.
15. Fan, C., & Ding, Y. (2019). Cooling load prediction and optimal operation of HVAC systems using a multiple nonlinear regression model. *Energy and Buildings*, 197, 7–17.
16. Shao, Y.L., Soh, K.Y., Wan, Y.D., Kumja, M., Zaw, K., Islam, M.R., Chua, K.J. (2020). Simulation and experimental study of thermal storage systems for district cooling system under commercial operating conditions. *Energy*, 203, 117781.
17. Zaw, K., Matsuoka, K., & Poh, T.K. (2023). A real-time operational optimization of commercial district cooling systems: A practical application. *Energy and Buildings*, 297, 113434. https://doi.org/10.1016/j.enbuild.2023.113434
18. Shipe, M. E., Deppen, S.A., Farjah, F., & Grogan, E.L. (2019). Developing prediction models for clinical use using logistic regression: An overview. *J Thorac Dis.*, 11, 574–584. https://doi.org/10.21037/jtd.2019.01.25
19. Yu, F.W., Ho, W.T., & Wong, C.F.J. (2024). Improved energy management of chiller system with AI-based regression. *Applied Soft Computing*, 150, 111091. https://doi.org/10.1016/j.asoc.2023.111091
20. Ho, W.T., & Yu, F.W. (2021). Prediction chiller system performance using ARIMA-regression models. *Journal of Building Engineering*, 33, 101871. https://doi.org/10.1016/j.jobe.2020.101871
21. Bhattacharjee, P., Dey, V., & Mandal, U. K. (2020). Risk assessment by failure mode and effects analysis (FMES) using an interval number based logistic regression model. *Safety Science*, 132, 104967. https://doi.org/10.1016/j.ssci.2020.104967
22. Zaw, K., Liu, G.R.. Deng, B., & Tan, K. B. C. (2009). Rapid identification of elastic modulus of the interface tissue on dental implants surfaces using reduced-basis method and a neural network. *Journal of Biomechanics*, 42, 634–641.

23. Deng, B., Tan, K. B. C., Lu, Y., Zaw, K., Zhang, J., Liu, G. R., & Geng, J. P. (2009). Inverse identification of elastic modulus of dental implant-bone interfacial tissue using neural network and FEA model. *Inverse Problems in Science and Engineering*, 17, 1073–1083. https://doi.org/10.1080/17415970903063151
24. Ali, M., & Hussein, M. (2023). Characterization and optimization of mechanical properties in design materials using convolutional neural networks and particle swarm optimization. *Asian J Civ Eng*. https://doi.org/10.1007/s42107-023-00918-5
25. Alzubaidi, L., Zhang, J., Humaidi, A.J., Al-Dujali, A., Duan, Y., Al-Shamma, O., Santamaria, J., Fadhel, M.A., Al-Amidie, M., & Farhan, L. (2021). Review of deep learning: Concepts, CNN architectures, challenges, applications, future directions. *J Big Data*, 8, 53. https://doi.org/10.1186/s40537-021-00444-8
26. Chaabene, W.B., Flah, M., & Nehdi, M.L. (2020). Machine learning prediction of mechanical properties of concrete: Critical review. *Construction and Building Materials*, 260, 119889.
27. Zhang, X. S. (2010). *Neural Networks in Optimization*. Kluwer Academic Publishers.
28. Liu, G.R., Han, X., & Lam, K.Y. (2002). A combined genetic algorithm and nonlinear least square method for material characterization using elastic waves. *Comput. Methods Appl. Mech. Engrg.*, 191, 1909–1921.
29. Wu, Z. P., Liu, G. R., & Han, X. (2002). An inverse procedure for crack detection in anisotropic laminated plates using elastic waves. *Engrg. Comput.*, 18, 116–123.
30. Zeng, C., Zhang, Q., & Wei, X. (2012). GA-based global path planning for mobile robot employing A* algorithm. *Journal of Computers*, 7, 470–474.
31. Gad, A. G. (2022). Particle swarm optimization algorithm and its applications: A systematic review. *Arch Computat Methods Eng.*, 29, 2531–2561. https://doi.org/10.1007/s11831-021-09694-4
32. Zhang, L., Wang, S., Selezneva, M.S., & Neusypin, K. A. (2022). A new adaptive Kalman filter for navigation systems of carrier-based aircraft. *Chinese Journal of Aeronautics*, 35, 416–425. https://doi.org/10.1016/j.cja.2021.04.014
33. Lee, J.H., & Ricker, N.L. (1994). Extended Kalman filter based nonlinear model predictive control. *Ind. Eng. Chem. Res.*, 33, 1530–1541. https://doi.org/10.1021/ie00030a013
34. Rectifier (neural networks) – Wikipedia. https://en.wikipedia.org/wiki/Rectifier_(neural_networks)
35. Mansini, R., Ogryczak, W., & Speranza, M.G. (2015). *Linear and mixed integer programming for portfolio optimizarion*. EURO Advanced Tutorials on Operational Research, Springer International Publishing Switzerland.
36. SAS institute Inc. (2015). *SAS/OR® 14.1 User's guide: Mathematical programming*. SAS Institute Inc.
37. Sequential quadratic programming – Wikipedia. https://en.wikipedia.org/wiki/Sequential_quadratic_programming
38. Fletcher, R. (2000). *Practical methods of optimization*. John Wiley & Sons Ltd. https://doi.org/10.1002/9781118723203
39. Zhibin, L.V., Ding, H., Wang, L., & Zou, Q. (2021). A convolutional neural network using dinucleotide one-hot encoder for identifying DNA N6-methyladenine sites in the rice genome. *Neurocomputing*, 422, 214–221.

40. Zaw, K., Kwik, Z.Z.S., Chang, W.Q., Islam, M.R., & Poh, T.K. (2023). A techno-commercial decision support framework for optimal district cooling system design in tropical regions. *Applied Thermal Engineering*, 220, 119668.
41. Cabeza, L. F. (2015). *Advances in thermal energy storage systems: Methods and applications*. Elsevier Pte Ltd.
42. Gulich, J. F. (2020). *Centrifugal pumps*. Springer.
43. Agarwal, R.. Patil, A., & Morrison, G. (2019). Efficiency prediction of centrifugal pump using the modified affinity laws. *Journal of Energy Resources Technology, Transactions of the ASME*, 142, 1–3. https://doi.org/10.1115/1.4044940
44. Coker, A. K. (2007). *Ludwig's applied process design for chemical and petrochemical plants incorporating process safety incidents* (4th ed.). Gulf Professional Publishing.
45. https://en.wikipedia.org/wiki/Root-mean-square_deviation
46. Singh, M., Dubey, R.K., & Kumar, S. (2022). Chapter-15 – Vehicle telematics: An internet of things and big data approach. *Artificial intelligence and machine learning for EDGE computing*. Academic Press. https://doi.org/10.1016/B978-0-12-824054-0.00019-8

6

District Cooling Systems: Benefits and Best Practices

6.1 Introduction

In Chapter 3, the intricacies of designing and developing district cooling systems (DCS) were discussed, offering readers a comprehensive understanding of how to create an efficient system during the development phase. In parallel, Chapter 4 provides a detailed financial analysis, aiding readers in making informed decisions regarding the development of DCS. Subsequently, as the DCS system becomes operational for providing cooling services to customers, Chapter 5 showcases real-time sustainable energy-efficient operations. However, a holistic grasp of DCS entails not only understanding its operational aspects but also recognizing its benefits and best practices. Therefore, this chapter is established in two folds. The first section compares conventional cooling systems (CCS) with DCS, highlighting the numerous benefits of the latter. This aims to provide readers with a clear understanding of the advantages offered by DCS. The second section focuses on delineating the best practices within the DCS industry, aimed at empowering readers with practical and theoretical knowledge to optimize the benefits of DCS development.

6.1.1 District Cooling versus Conventional Cooling System in Brief

Today, nearly all building types are outfitted with air conditioning systems. While various air conditioning systems exist, building cooling systems typically fall into two main categories: conventional cooling system (CCS) and DCS.

In a CCS, an individual building is equipped with its own cooling system to cater to its specific space cooling needs. Specially, a chiller system is installed and usually located in the basement while the cooling tower is seated on the rooftop for heat dissipation, see Figure 6.1(a). An advantage of CCS is that its implementation is more straightforward to execute. The CCS represents

DOI: 10.1201/9781032670164-6

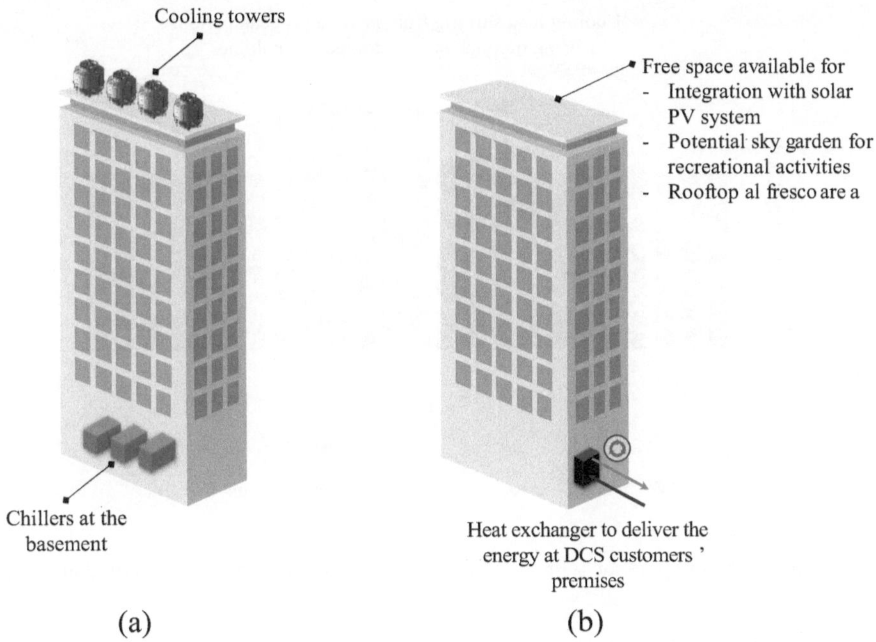

FIGURE 6.1
(a) Building with conventional cooling systems (CCS) with chillers at the basement and cooling towers on the roof of the building. (b) Building receiving chilled water from DCS through a heat exchanger in the basement [1].

a dedicated infrastructure while the DCS represents a shared infrastructure. Therefore, the CCS requires less initial planning and infrastructure design at the project development phase than DCS. The main challenge with CCS lies in managing each building's cooling facilities independently. In addition, the inability to capitalize on economies of scale becomes its significant drawback, as individual chiller plant operates independently within each building, leading to a lack of aggregation of cooling loads across multiple buildings in a district.

Contrary to CCS, buildings within the DCS zone do not need their own cooling equipment as chilled water is directly supplied by a centralized chiller plant. One of the most crucial components in the DCS is the thermal energy storage (TES) tank. During the off-peak hours or periods of low cooling demand, cooling energy is generated—a portion of it is channeled directly to the buildings to meet their cooling needs and the surplus is stored in the TES tank for later use during the peak hours. This process of storing the excess chilled water produced during the off-peak hours is known as "charging". This in turn leads to reductions in the peak cooling load of the chiller and the peak period electric power demand. A graphical representation of the

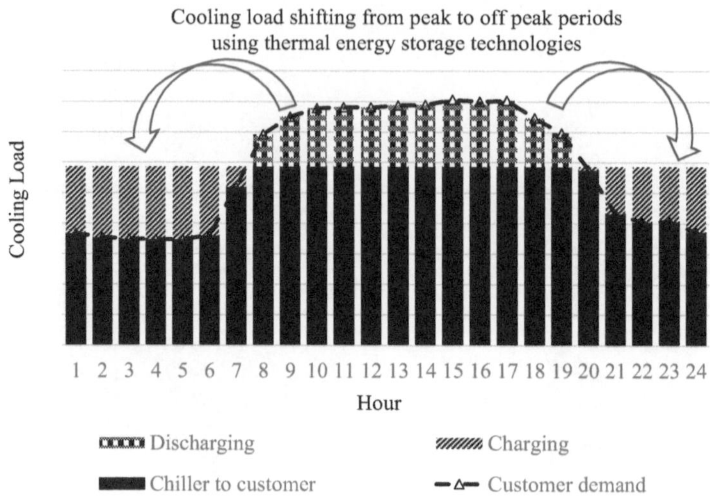

FIGURE 6.2
Load profile of chiller operation with thermal energy storage (TES) charging and discharging [1].

load-shifting technique by the TES is shown in Figure 6.2. In essence, the centralized chilled water production system coupled with the TES offers benefits to industry and society in a sustainable manner. Significant benefits of DCS naturally promote the widespread adoption of DCS technology.

6.2 Benefits of DCS

Numerous benefits offered by DCS are discussed in this section.

Benefit 1: The initial advantage is financial savings. Buildings located in DCS zones can function without the need for individual cooling systems. This results in substantial savings on capital investments, operational maintenance, and labor costs associated with this equipment [3]. With diligent monitoring and maintenance by skilled professionals, the overall system performance can be optimized, thereby reducing the life cycle cost of DCS as well [4]. From the perspective of DCS service providers, the integration of TES technology and cooling load aggregation plays a crucial role in generating both capital expenditure (CAPEX) and operational expenditure (OPEX) savings. The peak load shaving and shifting capability of TES along with cooling load aggregation allow for a lower chiller installed capacity in DCS compared to CCS. This results in several cost savings:

- Savings in operating costs due to differences in peak and off-peak electric energy tariffs.
- Savings in contracted capacity cost as DCS consistently utilizes less electric power compared to CCS, again due to load shaving and shifting (refer to Chapter 3 for detailed concept and Figure 6.2).
- Savings in maintenance cost due to the reduced chiller installed capacity.
- Cost savings due to improvements in energy efficiency (as described in Chapters 3 and 5).

These combined benefits translate into financial savings for both DCS operators and end users.

Benefit 2: Improvement in building aesthetics arising from space savings is the second benefit. With the absence of cooling energy generation facilities on building rooftops and basements, see Figure 6.1(a) and (b), space can be freed up to improve the aesthetics of the buildings, particularly for the DCS customers. Such architectural flexibility allows cooling towers on rooftops to be replaced with sky gardens, Al-fresco dining areas, terraces, and/or solar photovoltaic (PV) systems, while chillers in the basement can be replaced by underground pedestrian tunnels or carparks.

Benefit 3: The third benefit/offer is a reliable cooling service. With the presence of high-standard large-scale industrial equipment, 24/7 operation, and maintenance support that allows for an immediate response, DCS is able to offer more reliable cooling services as evidenced by its longer life span (i.e. 20–25 years) compared with CCS systems of typical mixed development (i.e. 10–15 years) [4]. In addition, cooling service disruption is also rare since DCS is built with backup cooling systems, TES tanks, and an emergency power generation unit that provides standby reserve. It is called as three-layer redundancy. A simple diagram is provided for a better understanding of the three-layer redundancy, see Figure 6.3. The first redundancy is backup or standby chiller that is usually in operation for emergency conditions like chiller breakdowns and annual chiller maintenance. The second redundancy is the TES tank. Its function is similar to the first redundancy. The last redundancy is an emergency in-house generator. It is dedicated to power outage conditions. In such a situation, the in-house generator is automatically in operation and supplies power to the distribution pump system in the DCS. The distribution pumps draw chilled water from the TES tank and supply chilled water to the customers. Such three-layer redundancy allows for the continuous supply of chilled water during regular maintenance, a chiller breakdown, and power outage, thus ensuring reliability.

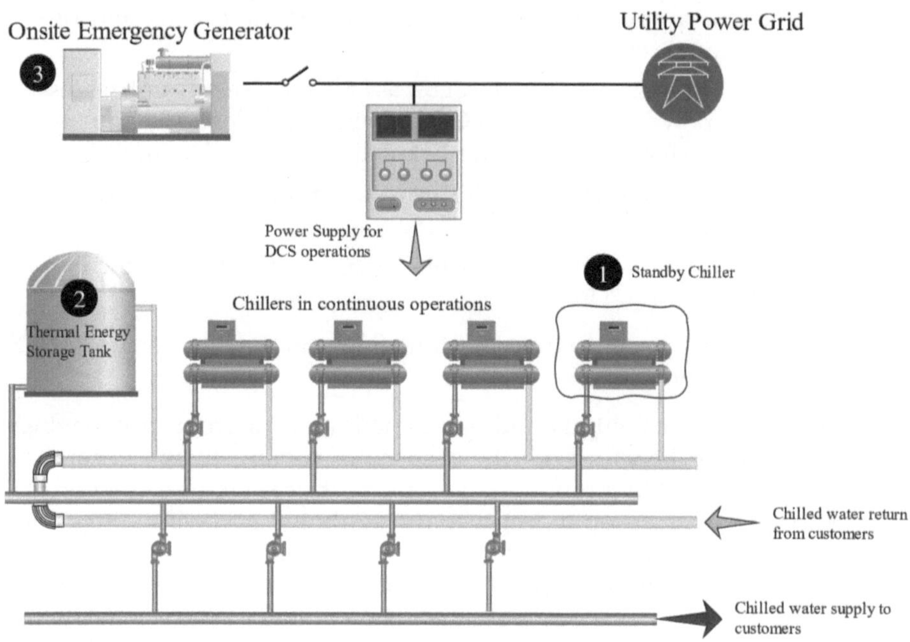

FIGURE 6.3
A typical DCS system with three-layer redundancies—(a) standby chiller, (b) thermal energy storage tank, and (c) onsite emergency generator—for improved service reliability.

Benefit 4: The fourth offer is flexibility, scalability, and economies of scale. DCS is flexible in that it allows the sharing of new and efficient chiller systems among all existing developments in the district, see Figure 6.4. This means even the oldest development benefits from the more efficient chillers. This is unlike the conventional CCS which is more rigid as any new chillers installed are dedicated only for a single building. Hence, the flexibility of DCS can benefit the entire district from technology advancement. Economies of scale is also created by aggregating the cooling needs of a cluster of buildings while satisfying variations in the cooling demand among these buildings. By aggregating the cooling energy loads and load shifting from the peak to off-peak hours with the aid of TES technology, the overall installed capacity of the chiller in DCS can be much lower than that of each building using a separate cooling unit or CCS [5], reducing embodied carbon emission associated with chiller manufacturing. Moreover, DCS creates opportunities to replace old inefficient chillers with new efficient chillers ahead of their economic lifespan as economies of scale makes the replacement of chillers justifiable commercially.

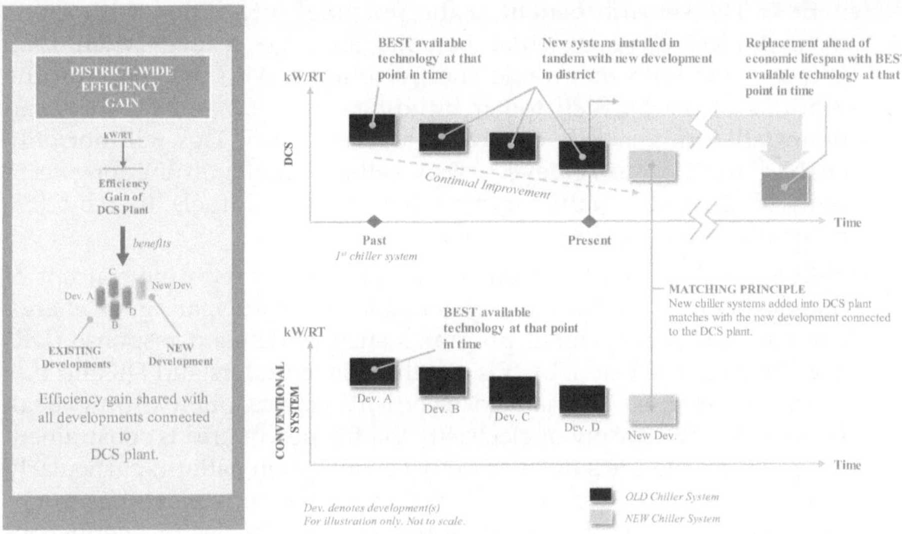

FIGURE 6.4
A DCS system with a phase development schedule, showing flexibility and scalability. (DCS installs a new chiller having better efficiency when the customer demand becomes higher. It shows scalability and flexibility.)

Benefit 5: The fifth offer is enhanced reliability of the electricity grid. For countries in hot climate zones, the increasing use of onsite CCS systems may stretch the strained electricity grid beyond its limits. However, with DCS, the storage of chilled water produced during the off-peak hours (night time) can help to reduce the pressure on the strained electricity grid while fulfilling the peak demand. Hence, with peak power demands on the grid avoided, pressure on the grid is reduced, and energy security is improved [6].

Benefit 6: The sixth is the reduction of noise, fuel consumption, and CO_2 emission. The DCS aids in reducing noise pollution in consumer buildings as the noise produced by cooling facilities is centralized and isolated away from individual buildings. As previously mentioned, with large-scale centralized chiller plants being more efficient than onsite small-capacity chiller plants [7], less electricity will be needed to run the chiller system for a given cooling demand. Furthermore, the ability to take advantage of the diversity of demand across all customers and implement technology like the TES allows a reduction of the peak cooling load production, and hence, the electricity required. With fuel consumption for electricity generation in power plants reduced, there will be less need for burning fossil fuels. As such, greenhouse gas (GHG) emissions such as carbon dioxide (CO_2) can also be reduced.

Benefit 7: The seventh benefit is the potential integration with green technologies. Environmental benefits are also reaped when DCS is integrated with renewable energy sources. With the space on the rooftops of the DCS customer buildings freed up, solar panels can be installed to generate electricity see Figure 6.5. This will not only enable buildings to achieve energy self-sufficiency in the long term but will also aid in reducing the reliance on fossil fuels, thereby dramatically reducing GHG emissions.

Benefit 8: Finally, the DCS is one of the pillars that form the concept of microgrid. DCS is particularly suitable for participating in energy demand-side management programs such as demand response (DR) and interruptible load (IL). It is worthwhile to understand DR and IL.

DR is designed to manage electricity consumption during peak hours when the supply of electricity via the power grid is constrained. DR program modifies the electricity consumption pattern, particularly during peak hours, through different incentive mechanisms, for example, time-based price incentive for the reduction of electric energy consumption, enhancing power grid stability, reliability, and overall power system cost-effectiveness. Another way of demand-side management is IL, which is a voluntary arrangement between electric energy service providers and energy consumers. In the IL program, the energy consumers or participants agree to reduce the electric energy consumption upon request by power grid operators/regulatory personnel. The IL program is designed to manage peak electricity demand to protect from potential power blackouts during periods when the utility grid tension is high. In other words, the IL program offers the balancing of supply and demand in the electricity market while mainly ensuring grid stability. The IL initiative offers financial incentives to the participants for reserving the reduction of electricity consumption. In both IL and DR activation periods, the participants are expected to curtail their electric energy consumption by different means, such as switching off non-critical load, switching to alternative energy sources, adjusting production schedules for manufacturers, or switching to backup power generation units. For the DCS, TES is an important technology that provides operation flexibility of operations by means of load shifting from peak to off-peak hours and operation rescheduling to fulfill the IL/DR requirement, see Figure 6.6.

The benefits of DCS are most pronounced in densely populated districts, since economies of scale can be achieved with high annual cooling demand load. This makes DCS extremely attractive in business cities given its densely populated area and high-density building clusters, coupled with its warm weather all year long that calls for greater cooling demand.

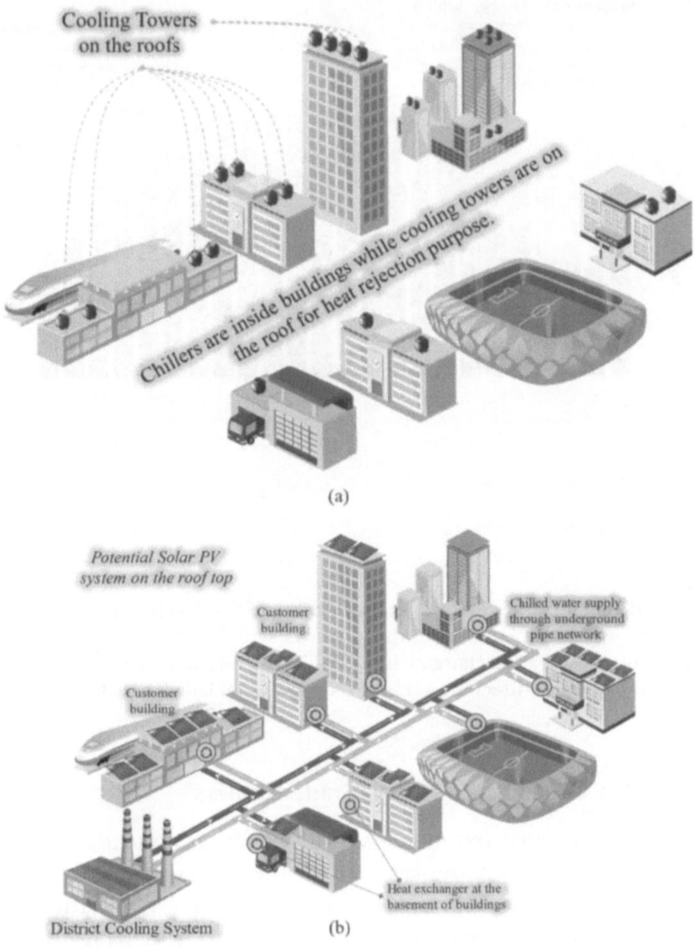

FIGURE 6.5
(a) Buildings with conventional cooling systems and (b) buildings receiving DCS services.

6.3 Best Practices

In the second part of the chapter, the best practices of DCS are discussed. The aims of this part are to summarize market available equipment related to the DCS system and selection and designing of this equipment. This part is composed of five different sections: (i) Refrigerant gases, (ii) chiller selections, (iii) TES system design, (iv) chilled water pipe network, and (v) testing and commissioning.

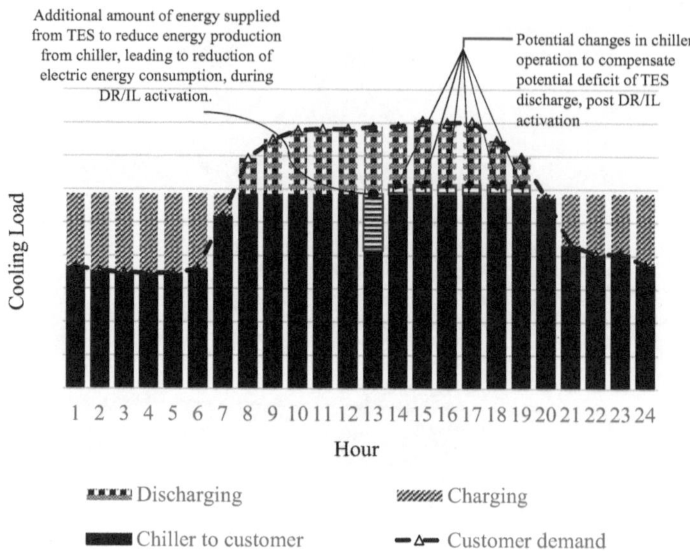

FIGURE 6.6
Changes of TES discharge and chiller operations due to DR/IL activation. During DR/IL activation, cooling energy production from the chiller is significantly reduced, but an additional amount of cooling energy is discharged from the TES while meeting cooling demand at the particular period. This has a direct impact on the operation of later hours after the DR/IL activation period.

6.3.1 Refrigerant Gases for Industry-Grade Chillers

Refrigerants, compounds extensively utilized in air conditioning and refrigeration applications, have evolved over the years. Initially, refrigeration applications relied on compounds like ammonia (NH_3), methyl chloride (CH_3CL), and sulfur dioxide (SO_2), until 1929. However, a significant turning point occurred in 1929 when fatal accidents linked to methyl chloride leaks spurred the quest for safer alternatives [8].

Extensive research and development efforts have led to the discovery of improved refrigerants, with advancements continuing up to the fourth generation currently available. These generations encompass first-generation chlorofluorocarbons (CFCs), second-generation hydrochlorofluorocarbons (HCFCs), third-generation hydrofluorocarbons (HFCs), and fourth-generation non-fluorinated refrigerants (HFOs), as of today.

Brief History of Refrigerants: The CFC compound comprising carbon, fluorine, and chlorine atoms is known as the first-generation refrigerant, invented in 1928. The first commercial CFC refrigerant, known as dichlorodifluromethane or R-12, was introduced in the 1930s [9]. Besides R-12, there are other chemical compounds within the CFC family. In the 1970s, scientists discovered that CFC compounds accumulating in the atmosphere had a detrimental effect on the ozone layer. When exposed to ultraviolet radiation, CFCs would break down, releasing chlorine atoms into the atmosphere and posing a significant threat to the ozone layer. As a result, an ozone hole was identified over Antarctica, allowing harmful ultraviolet radiation to penetrate the Earth's surface, leading to skin cancer and genetic damage in humans.

Hence, the second-generation refrigerants, known as hydrochlorofluorocarbons or HCFCs, were invented, designing less ozone depletion potential (ODP) and global warming potential (GWP) in comparison to CFCs. In addition, HCFCs have a shorter atmospheric lifespan than CFCs and exhibit greater susceptibility to natural breakdown due to their lower stability in the lower atmosphere. Nonetheless, it is important to note that HCFCs still contain a small amount of chlorine, which has negative implications for the ozone layer.

The third-generation refrigerants, referred to as hydrofluorocarbons or HFCs, were created with the aim of removing chlorine content from refrigerants to mitigate the detrimental effects on the ozone layer. Consequently, HFCs have found widespread application in various cooling systems across industries. However, a drawback of HFCs is their contribution to GHG emissions and their longer atmospheric lifespan compared to HCFCs. The extensive utilization of HFCs results in a significant increase in GHG concentration in the atmosphere, thereby exacerbating the effects of climate change. Ongoing research and development in refrigeration technology aim to explore potential new refrigerants boasting advanced properties and environmentally friendly attributes. This advancement could potentially be referred to as fourth-generation refrigerants, named HFOs. Other natural refrigerants, that is, CO_2 and ammonia, might be considered as the fourth-generation refrigerants, but it is crucial to stay updated with the latest information from experts.

Considering environmental impacts, briefly touching on GHGs, GWP, and ODP is worthwhile. GHGs include carbon dioxide (CO_2), water vapor, nitrogen monoxide (N_2O), methane (CH_4), and CFC compounds. These gases in the atmosphere scatter solar radiation, absorb, and trap energy, aiding in regulating atmospheric temperature. However, industrialization significantly increases GHG concentrations, leading to global warming, promoting environmental control measures. GWP quantifies a GHG's impact on global temperature, with CO_2 having a reference value of 1 (GWP of CO_2 = 1). Similarly, the ozone layer shields the Earth from harmful ultraviolet radiation. Increased UV radiation exposes risks to human, animal, and plant health, including skin cancer and eye diseases. ODP measures a GHG's impact on the ozone layer, referring to the first-generation refrigerant, named R-11, with a value of 1 (ODP of R-11 = 1).

Refrigerants and Environmental Impacts: Globally, the environmental issues originated from the consumption of fossil fuel due to the rapid growth of industrial development starting from the 1990s and the increasing development of the air conditioning industry have been discussed and measured, and corrective actions have been taken in a continuous manner. As a result, the Earth's temperature increased by >1°C since the 1880s, see Figure 6.7 showing global temperature in 1884, Figure 6.8 illustrating global temperature in 2022, and Figure 6.9 [10, 11] illustrating global temperature changes. The experts estimate that the Earth's temperature will be continuously increased. Global warming contributes to (i) a rise in sea level due to melting ice in the

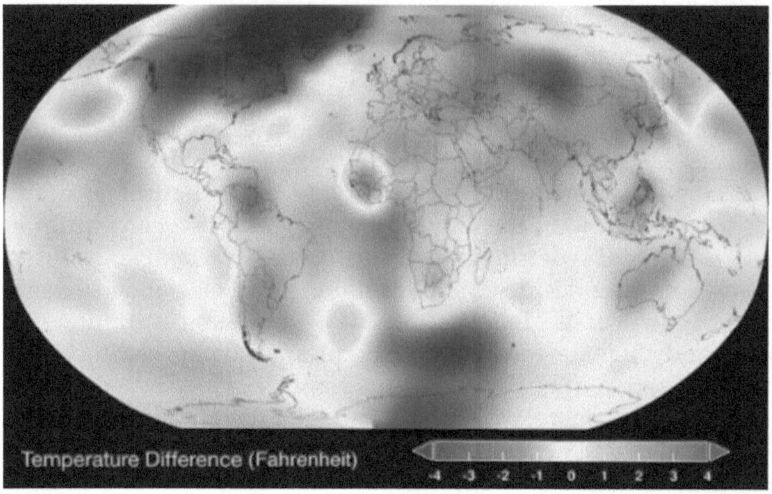

FIGURE 6.7
Global temperature map in 1884 [Ref. 10, Data source: NASA/GISS, Credit to NASA's Scientific Visualization Studio].

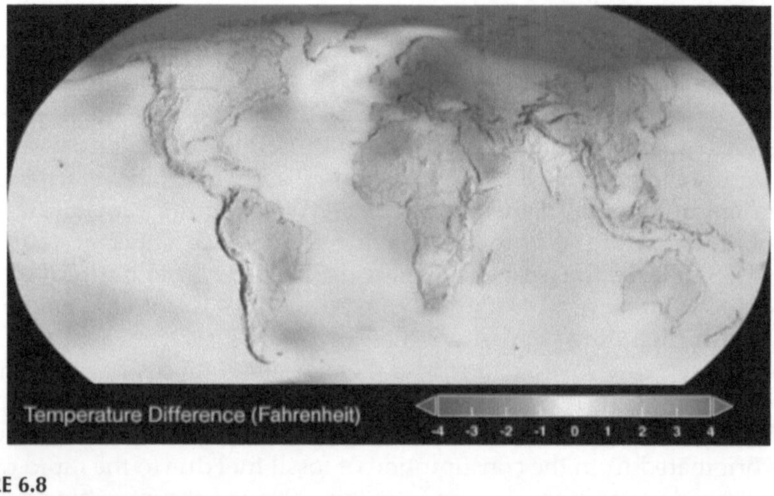

FIGURE 6.8
Global temperature map in 2022 [Ref. 10, Data source: NASA/GISS, Credit to NASA's Scientific Visualization Studio].

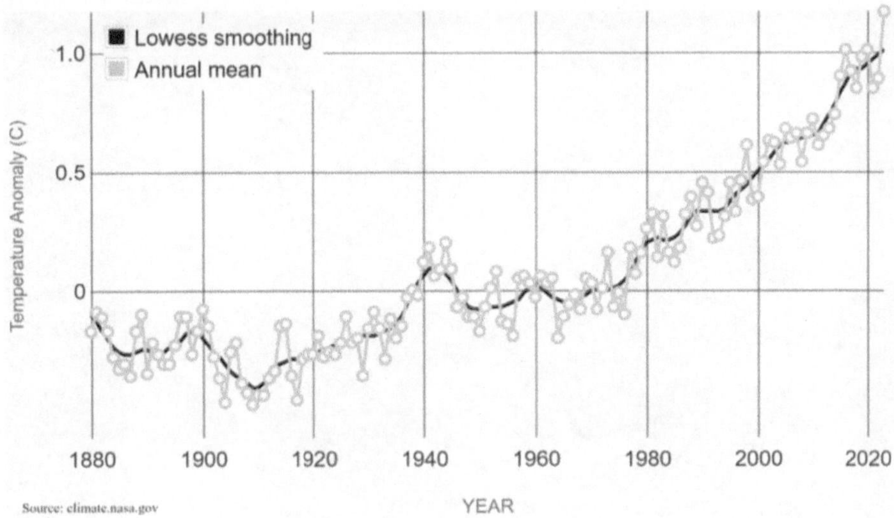

FIGURE 6.9
Global temperature changes from 1880 to 2023 [10].

Arctic and Antarctica regions and (ii) an increase in water evaporation from the oceans and seas, creating a favorable condition for forming storms.

In addition to global warming, ozone depletion was observed in the late 1970s, from the direct measurement over Antarctica. Factually, the Earth is enclosed by the ozone layer with an ozone concentration of ~300 Dobson units (DU). The DU represents the concentration of ozone [12]. One DU contains about 2.69×10^{16} ozone molecules—(i) for every 0.01 mm thickness at the standard condition of temperature (0°C) and pressure (1 atm) or (ii) for every 1 cm^2 of area at the base of the column. The Earth's ozone layer has an average thickness of 3 mm. Before 1979, the concentration of the Earth's ozone layer was always more than 220 DU. Therefore, the value 220 DU was defined as the baseline. In other words, the area of the ozone layer where the ozone concentration level is less than 220 DU is considered as the ozone loss potentially due to GHGs. The continuous monitoring of the ozone hole has been conducted from the time when the Antarctic ozone hole was found. From 1979 until now, the ozone hole has been measured. Figure 6.10 shows the ozone hole measured on every 1st of October from 1979 until 2021. Figure 6.11 shows a graph of DU values in different years, and the area of the ozone hole is given in Figure 6.12. The largest ozone hole occurred in 2006.

Refrigerator utilization was found to be one of the main causes of developing ozone holes. Thus, it is crucial to protect our environment sustainably while improving intelligently GWP and ODP. Since this book especially focuses on cooling applications, refrigerants used in cooling systems are discussed.

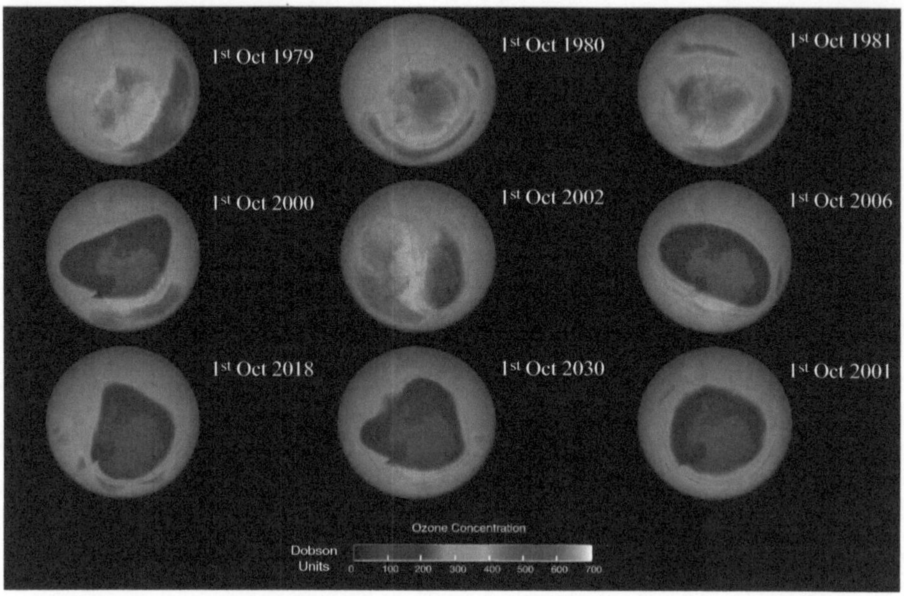

FIGURE 6.10
Illustration of ozone hole from 1979 until 2021 [13].

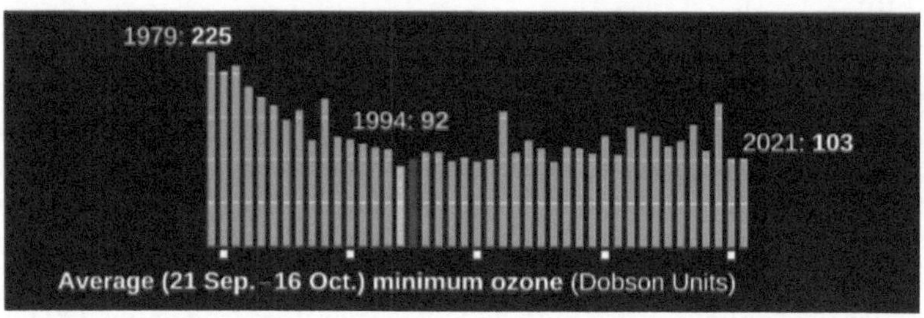

FIGURE 6.11
The value of Dobson Unit over Antarctica [13].

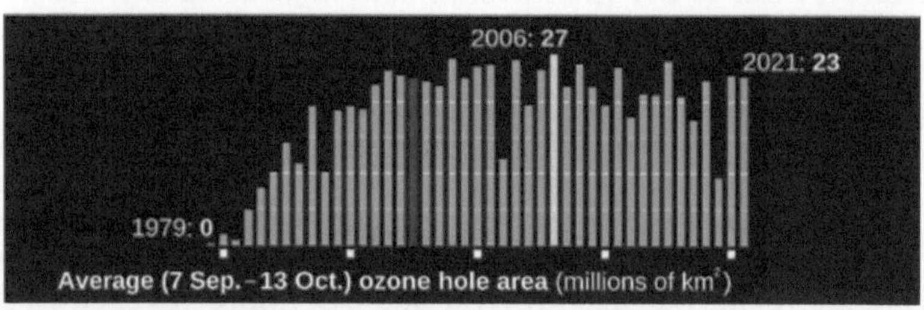

FIGURE 6.12
Area of ozone hole measured from 1979 until 2023 [13].

Market Available Refrigerants and Their Properties: Table 6.1 lists refrigerants that are commonly used for cooling applications (but not exhaustive) and their properties—ODP, GWP, toxicity, and atmospheric life span. Passing from the old generation to the latest generation, refrigerants have been systematically and successively developed to exhibit better environmentally friendly characteristics. To discuss prioritizing ODP and GWP, the first-generation refrigerants possess the highest ODP and GWP values, demonstrating notable environmental concerns. Therefore, second-generation refrigerants have been developed aiming at reducing the negative impact on our environment. A significant reduction in both ODP and GWP was found in the second-generation refrigerants. Continuous evolution introduced the transition to third-generation refrigerants, called HFCs. The HFCs effectively eliminate the ODP effect, though their GWP values still necessitate special attention. Continuous R&D efforts with environmental friendliness and sustainability in mind, the fourth-generation refrigerants which aim to eliminate or almost eliminate ODP while drastically lowering GWP, are developed and utilized. Of course, there are other refrigerants, that is, CO_2 and ammonia. However, these refrigerants have their own limitations in terms of energy efficiency and/or safety, leading to less utilization in cooling applications.

To deliberate safety classification of refrigerants, different safety classifications of refrigerants are given in Tables 6.1 and 6.2. It is realized that most of the refrigerants available in the market possess A1 grade, where "A" represents low toxicity and "1" refers to no flammability, except R-123 (safety class B1), R-32 (safety class A2L), R-1234yf (safety class A2L), and R-1234ze (safety class A2L).

TABLE 6.1

Different Generation Refrigerants and Their Properties

Refrigerants		Life Time	ODP	GWP	Toxicity	Flammable
CFC	R-11	45	1	4750	A	1
	R-12	100	0.82	10900	A	1
HCFC	R-22	11.9	0.034	1760	A	1
	R-123	1.3	0.012	79	B	1
HFCs	R-134a	13.4	0	1300	A	1
	R-410A	28.2	0	1924	A	1
HFC/HFO	R-32	5.2	0	677	A	2L
	R513A	5.9	0	572	A	1
HFO	R-1233zd (E)	0.0712	0.0002	1	A	1
	R-1234yf	0.031	0	1	A	2L
	R-1234ze (E)	0.038	0	1	A	2L
Natural refrigerants	CO_2	50	0	1	A	1
	Ammonia	0.01	0	0	B	2L

Ref: [1003525581-1003525584 - HPAC MagazineHPAC Magazine]

TABLE 6.2
Safety Classification for Refrigerant Gases

	Lower Toxicity	Higher Toxicity
Higher flammability	A3	B3
Lower flammability	A2	B2
Lower flammability	A2L	B2L
No flame propagation	A1	B1

Ref: [1003525581-1003525584 - HPAC MagazineHPAC Magazine]

6.3.2 Global Effort for Transition of Old Generation to New Generation Refrigerants

Recognizing the significant depletion of the ozone layer, the Montreal Protocol was agreed internationally to protect the ozone layer by phasing out the implementation and production of ozone-depleting substances on 26 August 1987. As of now, the Montreal Protocol has been revised nine times—in 1990 in London, in 1991 in Nairobi, in 1992 in Copenhagen, in 1993 in Bangkok, in 1995 in Vienna, in 1997 in Montreal, in 1998 in Australia, in 1999 in Beijing, and in 2016 in Kigali—and has evolved [14, 15], aiming to recover the ozone hole in a systematic manner.

As a collective effort of industry, the first-generation refrigerants have completely been phased out by 2010 in both developed and developing countries. The phase-out plan of the second-generation refrigerants or HCFCs was mandated. According to the Montreal Protocol and Figure 6.13, the phase out was scheduled in two-folds. First for the developed countries, the complete phase out of HCFCs was reached by 2020. Second, for the developing countries, the phase-out schedule was defined in stages in the period of 2015 to 2030 (reduction of 10% by 2015, 35% by 2020, and 100% by 2030, compared to the level of HCFCs utilization in 2013). The HCFCs have been substituted by the third-generation refrigerants HFCs that do not contribute to ODP but possess high GWP. Therefore, the focus shifted to cater not only to ODP but also to GWP in the use of refrigerants. In 2016, under the Kigali Agreement to the Montreal Protocol, the phase-down plan for the third-generation refrigerants HFCs was imposed [16]. In general, the Kigali Agreement aimed to phase down the production and utilization of HFCs up to 85% approximately over the next 30 years. The fourth-generation refrigerants, including HFOs and natural refrigerants, having (almost) zero ODP and significantly lower GWP, gain special traction.

Due to the collective and global efforts of the research environment, industry push, and regulatory pull, several refrigerants are only used in the market with careful consideration. Currently, the most widely used refrigerants in commercial applications are R123 and R134a in chillers and industrial cooling applications. There are also other refrigerants— (i) R404A and R507

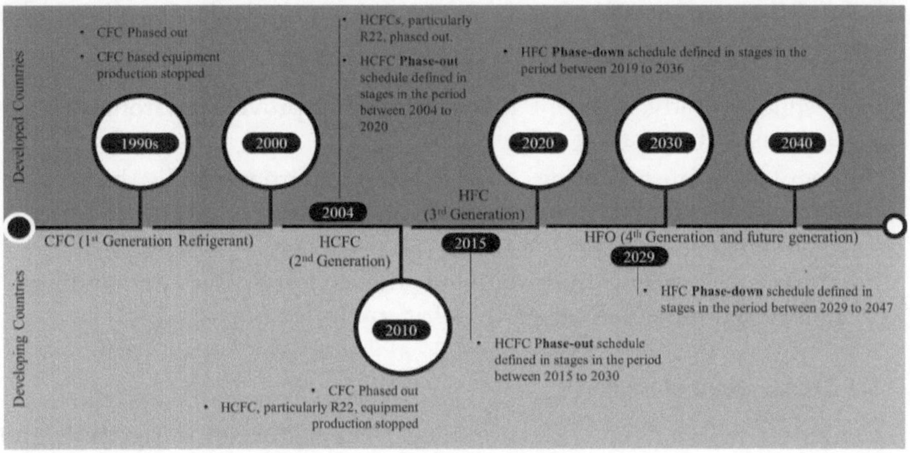

FIGURE 6.13
Different generations of refrigerants, phasing out of HCFC and phasing down of HFC.

for freezers and (ii) R290 for refrigerators. However, this book is dedicated for DCS, and thus, R123 and R134a will be discussed.

R123 is the second-generation refrigerant, with a low ODP of 0.012 and a lower GWP of 79. It is still in use in the cooling industry due to several reasons.

- Many existing cooling plants have been using the R-123-based chillers and are still in their economic life.
- R123 is still under a gradual phase-out schedule.
- R123 contributes a much lower GWP compared to R134a, R11, and R12. It might be a preferable choice while targeting for a balanced consideration of both GWP and ODP.

However, the use of R123 in chillers shall eventually be phased out.

Similar to R123, refrigerant R134a is commonly used in cooling applications, too. The refrigerant R134a possesses a ZERO ODP and a GWP of 1300. The main reasons for the continued use of R134a are as follows.

- R134a does not contribute to ozone depletion.
- Many existing cooling systems are featured with the R134a-based chillers.
- R134a, as a third-generation refrigerant, is in the phase-down phase, not in the phase-out phase.
- R134a is an economically viable and globally available option while transiting to more environmentally friendly refrigerants.

It is noted that the transition in refrigerants heavily dependent on three influential factors:

i. Technology advancement that can offer improved environmental impacts and operating efficiencies.
ii. Regulatory push aiming to phase out environmentally harmful refrigerants and promote the use of less harmful refrigerants under proposed local and international regulatory standards.
iii. Market pull arising from the well-educated end users demanding more environmentally friendly refrigerants.

6.3.3 DCS System Design

An engineer with a deep understanding of the history of refrigerants and environmental concerns should practice meticulous care when designing and selecting chiller systems for DCS. There are various standards to ensure the selection of chiller systems that meet up-to-date efficiency requirements. Notably, the Green Mark Standard [17] from the Building and Construction Authority of Singapore, the LEED standard [18, 19] and ASHRAE standard [20] are internationally recognized benchmarks. It is advisable for engineers to familiarize with these standards for comprehensive guidance.

In this section, sufficient information on designing a typical DCS plant while considering international standards is provided. In addition, practical rules of thumb will be offered. It is essential for engineers to apply this design guide in conjunction with the appropriate standards. The following topics will be covered in detail:

- Chiller selection
- Cooling tower selection
- Therma energy storage design
- Chilled water distribution networks
- Chiller plant room with control and monitoring systems

By integrating these guidelines and standards, engineers can ensure the efficient and environmentally responsible design of DCS plants. However, (almost) all chiller plants are designed to meet the ASHRAE efficiency requirement [21], see Figure 6.14.

6.3.3.1 Design Principle

Practically, cooling plants vary significantly from one another, primarily due to differences in the plant designer and consultant, the operational nature of

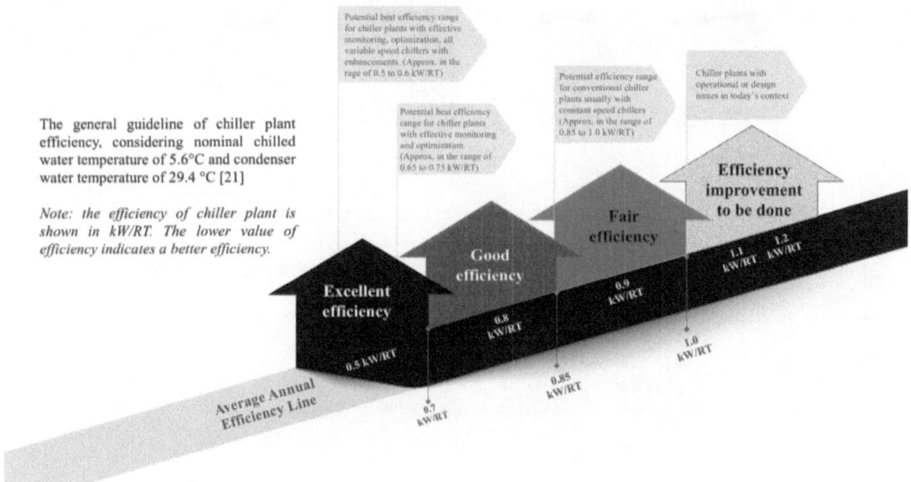

FIGURE 6.14
Classification of chiller plant efficiency generated from ASHRAE efficiency requirement and industry know-how.

the facility, and financial considerations. Each factor plays a crucial role in establishing the cooling plant.

First, individual chiller plant designers/consultants bring their unique expertise, experience, and preferences to the table. The individual quality leads to variation in the choice of chiller, chiller plant layout, and overall system performance. Second, the operational nature of the plant is another potential influence in the cooling plant design. The cooling load profile of end users necessitates tailored cooling plant design. Finally, financial considerations become imperative in the design and implementation stages. Budget constraints are one of the greatest challenges in most of the projects, resulting in variations in the selection of equipment and operational efficiencies.

Practically, these parameters collectively and significantly influence the design, evaluation, and implementation of cooling plants, which includes chillers, pumps, cooling towers, pipe network, and TES tanks, as illustrated in Figure 6.15.

6.3.3.2 Chiller and Pump Selection

First, chiller selection is discussed in stages.

Load Profile Analysis: During the design phase, analyzing cooling load profile is inevitable. The cooling load profile can vary based on the operational nature of the building. For instance, office buildings experience high cooling demand during office hours, whereas data centers

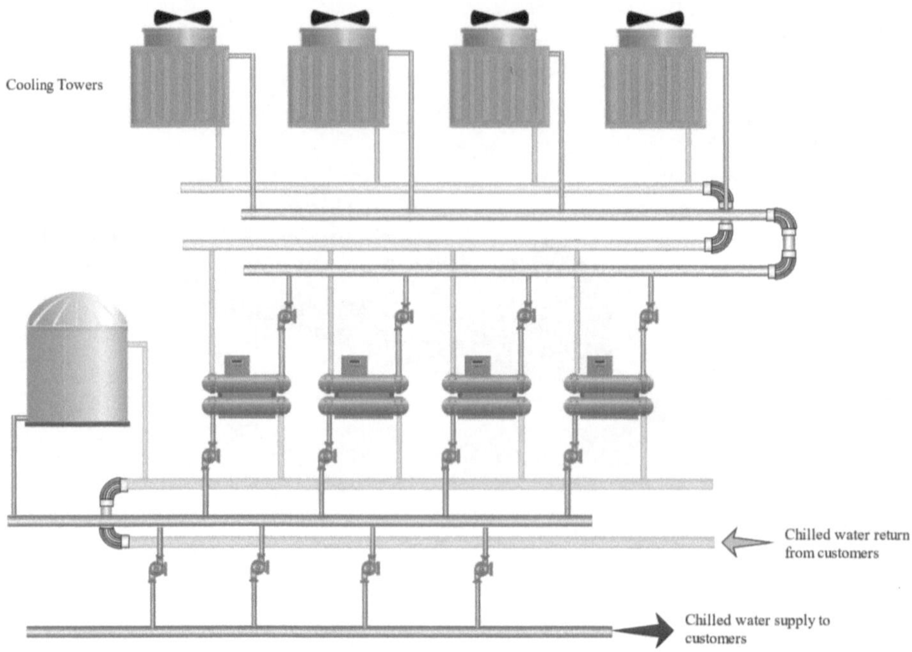

FIGURE 6.15
A typical DCS system and its components.

have nearly constant cooling demand throughout the 24-hour period. Therefore, conducting a cooling load profile analysis considering energy use patterns of commercial and industrial (C&I) buildings, research facilities, and other non-residential buildings is essential. The analysis will contribute to the appropriate sizing of the chiller plant before conducting an in-depth analysis of chiller efficiency and the selection of specific chillers.

Sizing and Efficiency: Based on the load profile analysis, the installed capacity of the chiller plant has to be defined, such as the capacity of the chiller and the number of chillers to be installed. In parallel, the efficiency of various market-available chiller options should be evaluated. Different chiller manufacturers offer varying chiller technologies with respective efficiencies. The performance data, usually expressed in kilowatts per refrigeration ton (kW/RT) or coefficient of performance (COP), along with a life cycle cost analysis, should guide the selection of chillers.

Key Considerations of Chiller Selection: The following key highlights are crucial for chiller selection.

- ***Capacity of Chiller***: Different capacities may necessitate the use of different compressor technologies [23], each with unique efficiencies and cost implications.

- ***Water-Cooled versus Air-Cooled Systems***: Water-cooled chiller systems use water from a cooling tower to transfer heat from the chiller to the atmosphere through evaporation. These systems typically offer better efficiency but require more space and have higher installation cost. Air-cooled system dissipates heat directly into the atmosphere via air movement. While less efficient than water-cooled systems, they are easier to install and require less space.
- ***Choice of Refrigerant***: Selecting the right refrigerant is imperative for efficient and sustainable operations. The selected refrigerant should comply with international protocols like the Montreal Protocol, which aims to reduce GWP and ODP, as discussed in Section 6.3.1.
- ***Performance of Chiller***: Chiller efficiencies are highly correlated to the capacity of the chiller, the use of refrigerant, the operating condition tailored to the specific needs, adherence to local/international regulatory standards and financial constraints. It is important to understand the chiller's performance at partial loads as chillers hardly operate at full load continuously. Integrated part-load value (IPLV) or non-standard part-load value (NPLV) are used to evaluate part-load efficiency [20, 22].
- ***Chiller Operating Conditions***: During the design stage of a chiller plant, understanding the local climate becomes the task of the design engineer, as local weather temperature range and humidity level affect the performance of the chiller. With such understanding and specific applications, design and operating temperatures for the evaporator (chilled water supply and return temperatures) and the condenser (condenser water supply and return temperatures) should be defined. For more detailed information on chiller operation conditions, the AHRI standard should be treated as a reference. For DCS and air-conditioning applications, the temperature difference between chilled water supply and return temperature, ΔT, is commonly in the range of 5°C to 7°C. Ideally, the higher ΔT improves the overall chiller system efficiency because the high ΔT allows for a reduced chilled water flow, leading to (i) a reduction of energy consumption of the pumps and (ii) a smaller chilled water pipe size and pumps for cost savings.
- ***Financial Constraint Versus Energy Efficiency***: The chiller should be selected at an optimal balance between financial constraints and energy efficiency while also ensuring compliance with local or international chiller standards.

Key Consideration of Pump Selection: The pump selection should be based on the cooling demand. In a DCS plant, multiple chiller systems are integrated as demonstrated in Figure 6.15. Each chiller system consists of a chiller, a primary chilled water pump, a condenser water pump, and a cooling tower. The primary chilled water pump is functioning for circulation of the chilled water between the evaporator of the chiller and a chilled water common header pipe, ensuring efficient and reliable delivery of chilled water to the secondary chilled water pump.

- *Pump Selection* requires a thorough understanding of the desirable water flow rate and the necessary pressure head/total pressure head. The flow rate determines the volume of fluid that needs to be flown. The total pressure head encompasses the static head, friction losses in pipes, and additional pressure losses due to fittings. The design pump pressure head must be sufficient to overcome the total pressure head requirements, ensuring the fluid is adequately pressurized throughout the system. Using these two important parameters, the pump size, with its performance and efficiency, can be determined based on the manufacturer's specifications. For a comprehensive understanding, readers should refer to this section alongside with Chapter 2.
- *Series Versus Parallel Pump Arrangement:* Depending on the applications, pumps can be arranged in series or in parallel. A series arrangement increases the pressure while maintaining the same flow rate. Conversely, a parallel arrangement increases the flow rate while maintaining the same pressure head. In DCS applications, it is crucial to have sufficiently high pressure to overcome the total pressure head and deliver chilled water to customers. Since customer demand often fluctuates, leading to changes in chilled water flow rates while maintaining the operational pressure head, secondary pumps in DCS are typically arranged in parallel. This setup ensures that the whole chilled water distribution network can adapt to varying flow requirements without compromising pressure stability.
- *Primary Chilled Water Pump Selection:* Understanding the required chilled water flow rate and the necessary pressure head for the connected chiller are two primary parameters for primary chilled water pump selection. It is worthwhile to note that each chiller has (i) its minimum chilled water flow rate that can be found in the chiller's specifications and (ii) maximum flow rate that should fulfill the full cooling capacity of the chiller.

The chiller will experience surging if the chilled water flow falls below its minimum flow rate. The required pressure head can be calculated based on the chilled water pump arrangement specific to the chiller system and the pressure loss in the evaporator.

- ***Condenser Water Pump Selection:*** Similar to the primary chilled water pump, the condenser water pump connected to the specific chiller can be designed. The main difference lies in the calculation of the required pressure heat. For the condenser water pump, it is important to account for the losses due to the pipe connection between the chiller's condenser and the cooling tower, as well as the static height difference between the chiller and the cooling tower area (if any).
- ***Secondary Pump System***: It should be designed to meet the cooling demand by ensuring, again, the chilled water flow rate and the necessary pressure head cover all pressure losses in straight pipe sections, fittings, strainers, and heat exchangers within the chilled water distribution network in the district. Unlike the primary and condenser water pump, the secondary pump system is ideally designed to maintain the system pressure but vary the chilled water flow rate that is heavily dependent on the cooling demand in real time. To achieve this, multiple pumps are arranged in parallel in the secondary pump system.

By carefully considering these factors, engineers can design and select the most appropriate chiller and pumps, ensuring optimal performance, energy efficiency, and sustainability.

6.3.3.3 ***Cooling Tower Selection***

In the cooling industry, vapor compression-based refrigerant cooling systems generate heat that must be dissipated through cooling towers. As discussed in Section 6.3.3.2, there are two different types of cooling towers: wet cooling towers for water-cooled chiller systems and dry cooling towers for air-cooled chiller systems. In the dry cooling tower, heat dissipation is accomplished by transferring heat from the condenser water to the atmosphere using a heat exchanger, see Figure 6.16.

In wet cooling towers, a combined heat and mass transfer method is employed for heat dissipation. High temperature water from the chiller is distributed in the cooling tower and sprayed onto the fill materials, as illustrated in Figure 6.17. These fill materials, known as infills, are designed to enhance heat exchange between air and water by increasing the surface area over the water spreads, thereby promoting better evaporative cooling. The infills are typically made from plastic and come in various configurations.

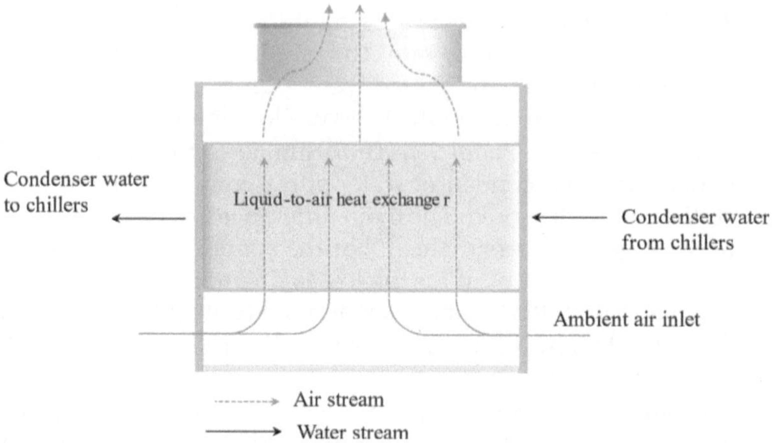

FIGURE 6.16
A schematic diagram of an air-cooled cooling tower.

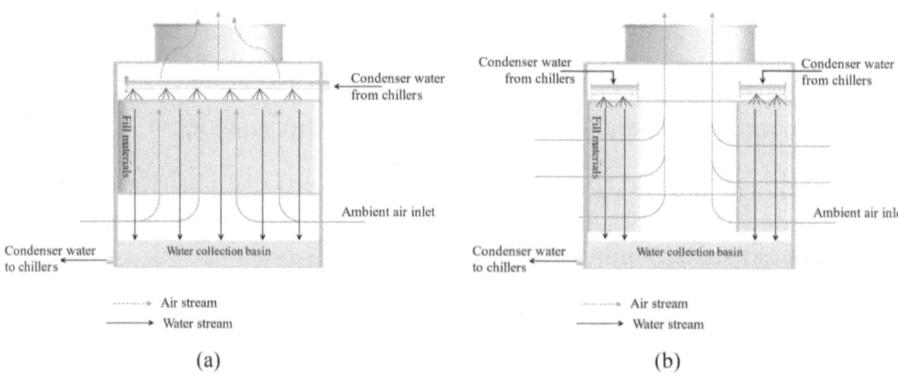

FIGURE 6.17
Schematic diagrams of water-cooled cooling towers: (a) counter-flow arrangement and (b) cross-flow arrangement.

Wet cooling towers can be categorized based on the arrangement of water and air flow: (a) counter-flow and (b) cross-flow cooling towers, as illustrated in Figure 6.17. In a counter-flow cooling tower, air flows vertically against the downward flow of water, resulting in more effective heat transfer due to longer contact time between the air and water streams. However, this design requires higher fan power consumption.

In contrast, a cross-flow cooling tower features horizontal airflow across a downward flow of water. This arrangement generally results in less effective heat transfer but lower fan power consumption than the counter-flow design.

Additionally, cross-flow cooling towers tend to operate at lower noise levels than counter-flow cooling towers. Noise level is a crucial parameter to consider when selecting a cooling tower, as different regions of the world have various environmental regulations, and noise pollution is a significant environmental concern.

Wet cooling towers are widely adopted by DCS due to their significant efficiency improvement for chiller systems. This is crucial in DCS because even a small efficiency improvement can lead to substantial annual operating cost savings due to economies of scale. In this section, the practical aspects of selecting cooling towers are discussed.

The following key considerations are important to consider when selecting and arranging cooling towers for DCS:

- **Performance Metrics**: Cooling towers should be designed to achieve an optimal approach temperature. The specifications of cooling towers are usually available.
- **Location and Air Circulation**: The area designated for cooling towers, known as the cooling tower yard, should facilitate good air circulation. To ensure effective heat dissipation, there should be no obstructions above the cooling towers. The prevailing wind direction and the stack effect of nearby structures shall be considered during the design stage.
- **Proximity Considerations**: The cooling tower yard should be at least 5 m away for air circulation and ventilation inlets, kitchen exhaust outlets, occupied areas, noise-sensitive areas, pedestrian thoroughfares, and trafficable areas.
- **Air flow simulation**: Simulation should be conducted to ensure proper airflow within the cooling tower yard, preventing cooling towers from competing for air. Proper arrangements should be made with the help of simulations and manufacturer suggestions. Ideally, the cooling tower yard should have at least 0.2 m^2 per refrigeration ton (RT), with towers arranged evenly.
- **Sizing**: Cooling towers should be sized as large as possible, given the available space, to achieve colder condenser water temperatures.
- **Variable Speed Drives**: Implementing variable speed drives on cooling tower fans is crucial for adjusting airflow and optimizing energy consumption based on cooling demand, wet-bulb temperature, and condenser water supply to the chiller.
- **Construction Materials**: The structure and components of cooling towers should be made from durable materials. For instance, fiberglass-reinforced polymer (FRP) is commonly used for its corrosion resistance, strength, and durability.

- **Noise Level**: Noise levels must be considered during the design stage to prevent noise pollution and comply with local/international environmental regulations.
- **Selection of Cooling Tower Type**: Depending on space availability, environmental concerns, and energy efficiency, either cross-flow or counter-flow cooling towers should be selected.

6.3.3.4 Thermal Energy Storage System

This section describes TES, a crucial part of cooling energy systems, for efficient utilization of cooling energy. TES technology offers:

- Storing cooling energy during off-peak hours when cooling demand is low, thereby utilizing cooling energy generation assets more effectively and efficiently.
- Discharging the stored cooling energy during peak hours to meet end-user needs, reducing the need for additional cooling energy production and effectively responding to the cooling demand in a real-time setting.

Technically, TES systems can be classified into two categories based on the energy storage method: (i) sensible energy and (ii) latent energy storage.

(i) Sensible Energy Storage: It is a well-known method with water being a commonly used medium for cooling energy storage due to its large storage capacity, low cost, and safe material properties. It is worthwhile to note that while other materials like bricks, stones, soil (or geothermal), and concrete are also used for heat energy storage, this book focuses on cooling energy storage. Water-TES tanks are widely deployed in the district cooling industry, with various types of tank structures employed. Figure 6.18 illustrates two main types of TES tanks—horizontal and vertical tanks. Horizontal water-TES tank usually lays horizontally and is installed in areas where more horizontal space is available. It can be either a cylindrical or rectangle tank. With an understanding of operations, specific needs, and constraints, the choice of thermal energy storage TES system shall be determined. However, horizontal water-TES is less effective in thermal stratification, leading to a significant impact on overall energy storage efficiency and reduced thermal storage capacity compared to vertical tanks. Multi-connected horizontal water-storage tanks (may) need to achieve the same TES as a single vertical water-TES. Therefore, vertical water-TES tanks are widely employed in the DCS industry due to their high energy storage efficiency and less space requirement. In this book, the vertical water-TES system is discussed.

Industry practices for integrating TES into chiller plants to form DCS systems are discussed. Key design and operational parameters are raised.

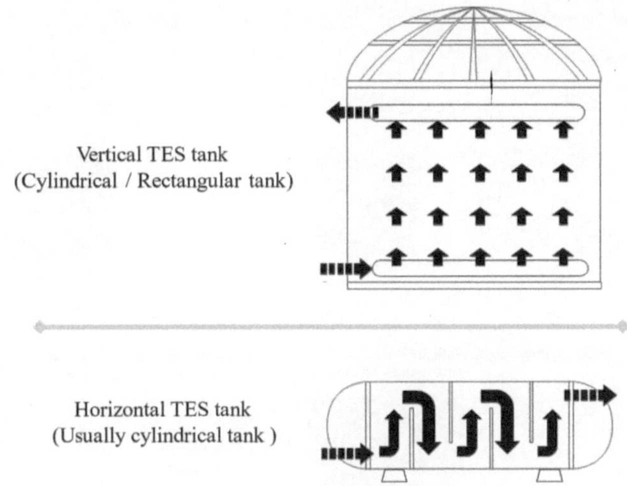

FIGURE 6.18
Vertical water thermal energy storage tank and horizontal water thermal energy storage tank.

- **Operation of Vertical Water-TES**: Understanding the charging and discharging procedures of a water-TES tank is essential. During charging, chilled water produced from the chiller (having a specific temperature, $T_{CHW,ECout}$) is pumped into the bottom of the tank gradually replacing the high-temperature water, $T_{TES,initial}$, in all levels of water inside the tank, until the tank is fully charged with cooling energy. During peak hours, the stored energy is discharged by releasing chilled water from the bottom while high-temperature water is pumped in from the top of the tank. Figure 6.19(a) shows the energy-depleted water-TES tank with high-temperature water $T_{TES,initial}$ at all levels of water inside the tank, indicating no useful cooling energy remains. Figure 6.19(b) and (c) illustrate the energy changing and discharging processes described.
- **Temperature Stratification and Chiller Operations**: To achieve optimal temperature stratification inside a water-TES tank, it is important to understand the process of cold energy charging. An example is explored to demonstrate the concept more clearly. Consider a vertical water-TES tank with a specific height, undergoing a charging process over six hours. When the chilled water distributors at the bottom and top of the tank are well-designed, the tank achieves effective temperature stratification. This stratification is crucial to prevent the mixing of cold and warm water, which would otherwise decrease the temperature difference between the outlet and inlet of the tank, reducing the efficiency of the chiller connected to the tank.

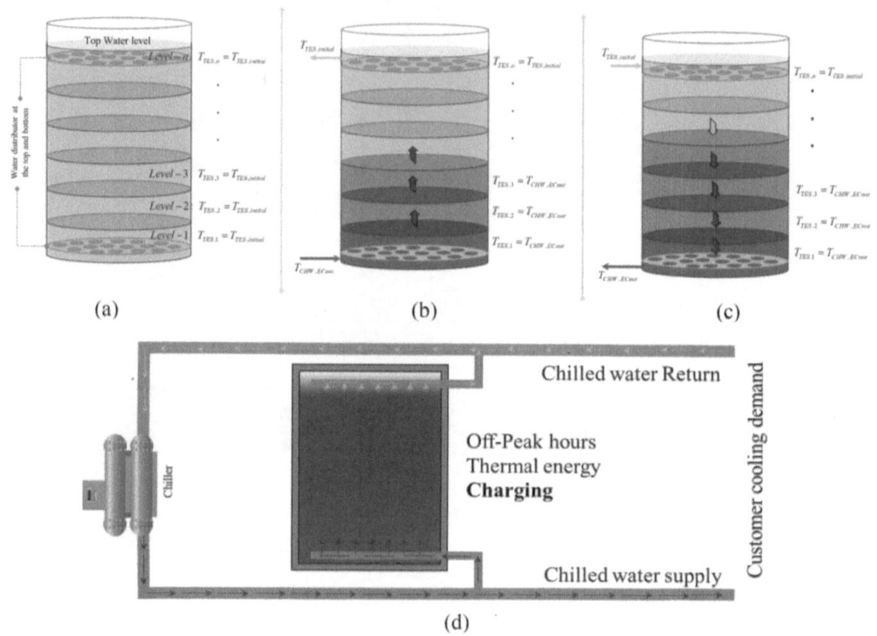

FIGURE 6.19
Water-based thermal energy storage tank (TES) with "n" layers of water: (a) TES with no useful energy; (b) charging energy into the TES; (c) discharging energy from the TES; (d) the integration of chiller and TES system in the DCS.

In this example, the tank is charged with 4°C chilled water into the energy-depleted tank with initial water temperature ranging from 12°C to 13°C. As the charging progresses, the temperature changes are observed as follows:

- After 1 hour: The temperature in the lower layers of the tank drops to between 6°C and 7°C up to level 3.
- After 3 hours: The temperature further decreases to between 5 and 6°C up to level 6.
- After 5 hours: The temperature stabilizes at 4.5°C to 5°C up to level 9.

Eventually, the entire tank's water temperature settles between 4.5°C and 5°C. This effective stratification, illustrated in Figure 6.20, demonstrates the benefits of a well-designed vertical water-TES tank, maintaining distinct temperature layers and ensuring efficient chiller operations.

- **Energy Storage Efficiency**: Energy efficiency of TES is a useful index in evaluating the heat loss of TES.

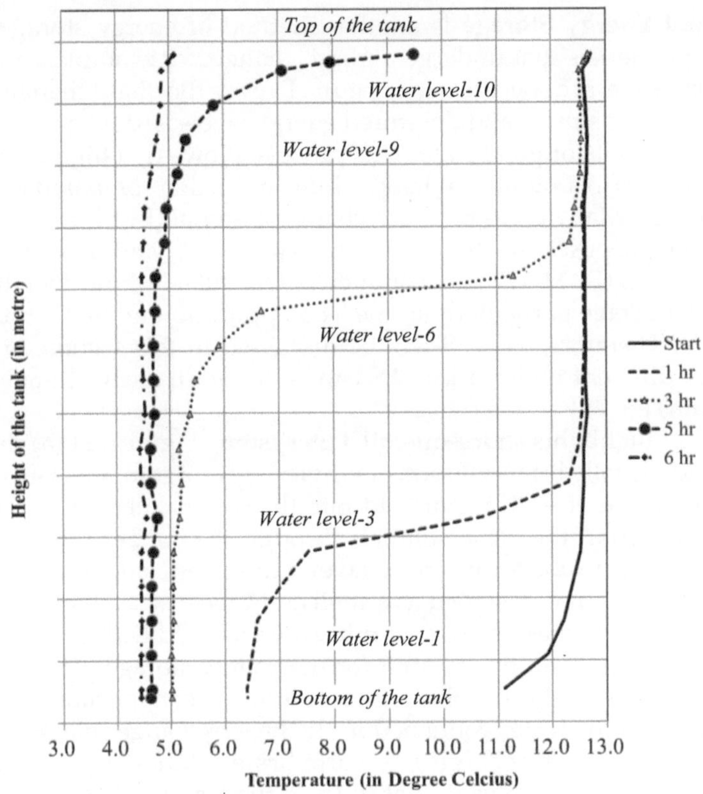

FIGURE 6.20
An illustration showing the water temperature inside the tank during charging period.

$$\eta_{TES} = \frac{Q_{stored}}{Q_{generated}}. \tag{6.1}$$

In Eq. (6.1), $Q_{generated}$ is cold energy generated from the chiller connected to the TES, and it can be simply calculated using the inlet and outlet temperatures of chilled water and its flow rate over the charging period. The term Q_{stored} is the stored energy inside the tank, and it can be estimated using the initial temperature of the water inside the energy-depleted TES tank, the final temperature of the water inside the TES tank after the charging period, and the total water volume inside the tank. The amount of thermal energy loss to the environment will vary depending on the type of insulation used on the tank walls. Typically, thermal energy loss could be around 5% for cooling applications. In addition to the tank's insulation, the installation and placement of temperature sensors inside the tank are also crucial factors for engineers.

(ii) Latent Energy Storage is another method of energy storage system employing materials that undergo a phase change, for example, from liquid state to solid state, at a specific temperature. During the phase change process, the cold energy is stored, and the stored energy is released when the material phase returns to its original state. This process allows (i) a high energy density storage and (ii) a significant tank volume reduction compared to sensible energy storage. Many types of phase-change-material (PCM) can be utilized for different purposes. In the DCS industry, ice-TES and a paraffin-based PCM-based TES (PCM-TES) are commercially available. However, the latter is in its infant stage particularly in the TES application in the DCS industry, as of today. Therefore, ice-TES will be discussed in this section. Similar to water-TES, different types of ice-TES can be found, namely, ice-on-coil TES and ice-ball TES.

Figure 6.21 highlights an **ice-on-coil TES system**. Figure 6.21 (a) shows the chilled fluid; usually brine solution or ethylene glycol solution at an approximate temperature of −5°C is pumped into the copper tube. Due to the heat transfer effect from the brine solution through the copper tube, the water around the copper tube forms an ice layer and stores cold energy. This type of ice storage system is designed as a shell and tube heat exchanger block. In the heat exchanger block, the copper tubes are arranged with proper spacing, and the two types of fluids are used for the energy storage process. During the charging process, brine solution flows into the copper tubes to transfer cold energy into the water contained in the heat exchanger block, as shown in Figure 6.21 (a). For discharging, a warm brine solution is passed through the tubes again to absorb cold energy from the ice, Figure 6.21 (b). In this type of TES, it is crucial to maintain proper copper tube spacing so that all the water in the heat exchanger block can form ice and achieve the designed cold energy storage capacity.

The Ice-Ball TES system, as shown in Figure 6.22, operates by placing water balls within a brine solution inside a tank. During the charging process, cold brine solution is introduced from the top of the tank, while warm brine solution is circulated back from the bottom to the chiller, as highlighted in Figure 6.22 (a). Due to the heat transfer theory, cold energy is transferred

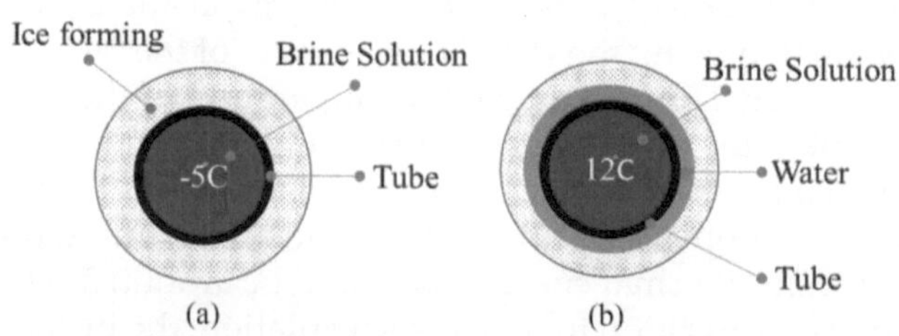

FIGURE 6.21
Ice-on-coil thermal energy storage.

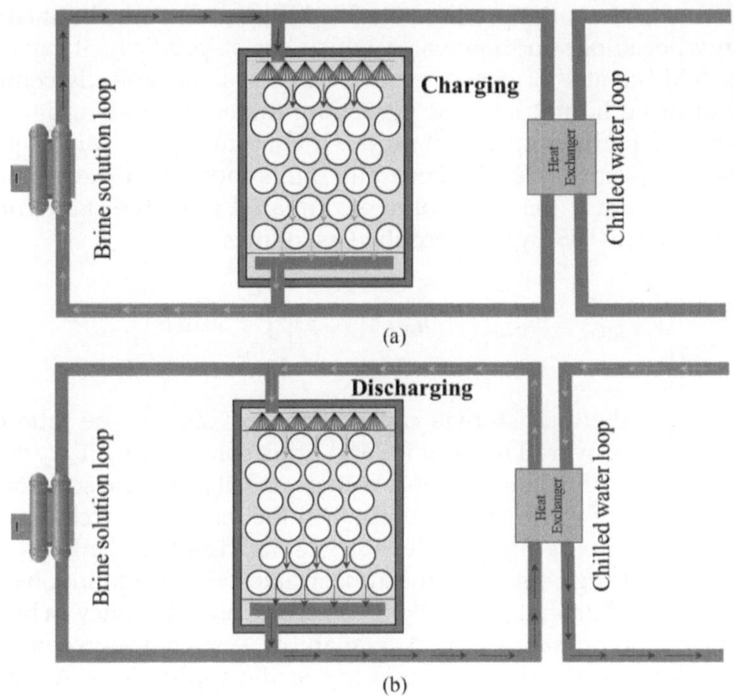

FIGURE 6.22
Ice-ball thermal energy storage integrated with chiller and heat exchanger: (a) energy charging process and (b) energy discharging process.

from the brine solution into the water through the plastic layer of the water balls. Notably, the tank contains very little brine solution, with most of the space occupied by the water balls. Therefore, during the charging process, the brine solution in the tank circulates quickly, unlike in a water-TES tank. This results in a cyclic decrease in the tank's temperature each hour, owing to the operation of brine or ice-making chillers. The brine solution continuously circulates, lowering the warm brine temperature to below freezing. Consequently, ice-making chillers consume significantly more energy than regular water-chilling chillers. During the discharging period, the ice-making chillers are inactive. Instead, the brine solution flows to the heat exchanger connected to the ice-ball TES tank, as shown in Figure 6.22 (b). This heat exchanger is essential because the DCS supplies chilled water to customers, not brine solution, due to the corrosive nature of brine.

The degree of ice storage tank volume reduction is largely influenced by the ice packing factor (IPF). The IPF, which represents the ratio of the volume of ice to the total volume of the TES tank, is a very important parameter in determining the effectiveness of ice-TES. To better understand this concept, consider the energy storage capacity of ice at 0°C and water at 4°C. Ice, with a latent heat of fusion of approximately 334 kJ/kg, can store about 80 times more energy per kilogram than water, which has a specific heat capacity of roughly 4.18 kJ/kg at 4°C. However, an ice-TES system typically comprises a mixture of brine liquid, ice, and packaging materials, making its energy storage capacity highly dependent on the IPF. From a practical design and operational perspective, ice-TES generally offers more than three times the cooling energy storage per unit volume compared to water-TES. Thus, the energy stored in ice-TES can be described as follows:

$$Q_{stored(ice)} = m_{Total}\left[\left(\gamma \times h_{sf}\right)+\left(1-\gamma\right)\left(\int_{T_{initial}}^{T_{final}} C_b dT\right)\right] \tag{6.2}$$

where $Q_{stored(ice)}$ is the total storage capacity of ice-TES, γ is the ratio of the mass of ice over the overall mass of ice and brine solution or IPF, h_{sf} is latent heat of fusion of ice, C_b is the energy-carrying capacity of brine solution, T_{final} is the temperature of the brine solution when the tank is fully charged, and $T_{initial}$ is the brine temperature of the energy-depleted storage tank.

In real-world settings, selecting the appropriate TES system involves considering various influential parameters. These include efficiency in both initial investment and operations, space availability, system design complexity, and commercial considerations tailored to specific applications. A thorough understanding of TES technologies, combined with practical experience in their application, can significantly aid in making informed decisions about TES technology deployment.

6.3.3.5 Chiller Plant Room Design

Beyond chiller and pump selection, the chiller plant room should be spacious enough for proper ventilation, servicing, and maintenance purposes. Based on the authors' experience, the following are highlighted:

- The spacious plant room with about 6 m high ceiling should be good for the chiller plant. Chiller plant space is allocated depending on the available space. However, the space for maintenance and chiller replacement activities should be properly allocated during the planning stage.
- To prevent the flood from potential pipe leakage and accidental discharge of sprinkler (as the chiller plant room is usually protected for fire safety), floor risk assessment shall be conducted during the design stage.
- The chiller plant should be able to supply continuous chilled water to critical service areas and vital facilities in the event of a power outage. Therefore, emergency generators shall be installed in the DCS plant.

- The refrigerant leak detection system, which is interlocked to the emergency purging ventilation system, should be provided to guard against the dangers of the refrigerant.
- The operating weight of the equipment and associated piping should be provided to the structural engineer during the concept design stage to inform the design of the plant room's substructure and superstructure.
- The plant room should be planned with future needs in mind, including potential expansions and ensuring sufficient access for equipment replacement from the initial design stage.
- The control and monitoring room should be integrated with the chiller plant room. It will house control panels and a human–machine interface (HMI) equipped with a graphical user interface, enabling operators to monitor and manage the chiller plant. The room will also feature an energy management system, alarm and notification system, data logging capabilities, and analysis tools.
- Electrical room usually occupies about 30% of the chiller plant room and houses various critical components, including:
    - Main electrical panel: Distributes electricity to the components of the chiller plant.
    - Variable speed drives (VSD): Control the speed of motors for chillers, pumps, and cooling tower fans,
    - Transformers: Step up/step down voltage from the utility grid to a lower voltage suitable for the equipment.
    - Switchgear: Includes circuit breakers, switches, and fuses that ensure safe operations and maintenance.
    - Automatic transfer switches: Transfer the electrical load from the primary power source to the backup generator/other power sources in the event of power failures.
    - Uninterruptible power supply (UPS) systems: Provide backup power to critical equipment during power failures.
    - Other auxiliary (electrical) systems: Support the energy management system, lighting, and other services.

The electrical room should be located above ground for safety, maintainability, and protection from environmental hazards.

6.3.3.6 Chilled Water Distribution Network

In the development of DCS, three water distribution networks are involved: the condenser water network, the chilled water network within the DCS premises, and the chilled water distribution network from the DCS plant to the end user.

The condenser water network, referred to here as the cooling water network, consists of a pipe connecting the chiller cluster in the DCS plant to the cooling towers. As shown in Figure 6.15, a common header is designed to receive cooling water pumped from the chillers. The water is then directed to the cooling towers for heat rejection. Engineers must design the diameter of the header pipe, branch pipes, elbows, and other necessary fittings. It is important to note that larger pipes allow for lower water velocity, which reduces friction loss and decreases pump power consumption. The chilled water network within the DCS premises can be designed similarly to the condenser water network. Pressure-independent control valves (PICV) should be deployed in these pipe networks to regulate and maintain the required flow set point. The PICV are usually integrated with a pressure regulator and adjust the valve position accordingly. Changes in pressure in the pipe network usually occur due to pump operations and energy demand variations. In a simple explanation, the PICV functions to maintain the system's required water flow by adjusting to pressure changes.

Lastly, the design of the chilled water distribution network must account for several factors: the cooling demand of DCS customers, the distance between the energy production source (the DCS plant) and the energy receiving point (the end user), the setup of heat exchangers at the energy-receiving stations, and the terrain of the DCS zone. Typically, DCS chiller plants are located in the basement, with the pipe network installed underground and customer energy-receiving station positioned in the basements of buildings within the DCS zone. This configuration ensures that the pump pressure reference point remains within the DCS plant, which is crucial for engineering purposes. The distribution pump pressure head must be designed to overcome flow resistance within the pipe network, including friction losses, dynamic losses, and losses due to fittings. As illustrated in Figure 6.23, the DCS distribution pump system maintains a minimum pressure head, H_p, which ensures the required chilled water flow and pressure, $H_{\text{allowable}}$, to the hydraulically farthest customer. To optimize energy consumption of the chilled water distribution pumps, the main distribution pipe should have a larger diameter

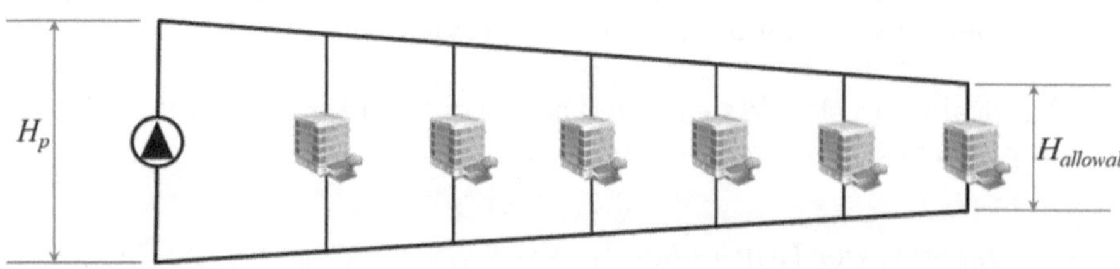

FIGURE 6.23
A simple pressure diagram of the DCS chilled water distribution network.

throughout the network, while pipe branches are sized according to individual customer cooling energy demand. However, it can be a hefty investment in the underground pipe network. A more economical option is to use a cascaded diameter design for the main distribution pipe where the section closest to the DCS plant is larger than the farthest section, as the latter carries a lower cooling energy.

6.4 Discussions

Designing an optimal DCS involves ensuring energy efficiency, sustainability, and reliability, as discussed in Section 6.3.3.1.

First and foremost, understanding the year-round cooling demand is essential for defining the design capacity of the chiller system, its auxiliary components, and TES. This foundation allows the chillers to operate at maximum efficiency consistently, supported by the TES.

Second, value-added engineering can be implemented across three aspects; (i) the selection of chiller capacity featuring sustainable and energy-efficient refrigerants, (ii) the choice of energy-efficient pumps, and (iii) the thoughtful design of cooling towers that adhere to, at a minimum, local or international energy efficiency standards, such as LEED [18, 19], ASHRAE [20], AHRI [22], and Singapore Standard 530 (SS530) [24]. These standards set minimum energy efficiency requirements for chillers, cooling towers, and pumps. For instance, the minimum chiller efficiency requirement is outlined in SS530 and AHRI standards. Therefore, with the aid of provided guidelines in this chapter and the above-mentioned minimum energy efficiency standards, DCS design and equipment selection should be carefully evaluated and balanced to develop an efficient, reliable, and cost-effective system tailored to the specific needs of the application.

Third, the configuration of the chiller plant room, control and monitoring station, and cooling tower yard significantly influences energy efficiency, accessibility, and maintainability. Optimal ceiling height in the chiller plant room and effective airflow in the cooling tower yard are critical. It is advisable to conduct simulation studies, especially in the area of the cooling tower, to optimize these elements.

Finally, to achieve optimal pump energy consumption—including the condenser and the chilled water pipe networks within the DCS premises, as well as the distribution network—it is crucial to design with larger pipe diameters, minimize fittings, and reduce the number of pipe bends. A blend of industrial expertise, operational experience, and fundamental engineering principles is essential for crafting an efficient water distribution network. Although the DCS plant is designed for energy-efficient operation, it is imperative to establish comprehensive operational and maintenance policies in several areas: (i) implementing an automatic tube cleaning system in the condenser water circuit of the chiller to maintain optimal heat transfer conditions; (ii) deploying effective water treatment systems to prevent scaling, corrosion, and biofouling in the chiller system and cooling towers; (iii) adopting a blow-down water recovery system to reduce the water consumption of cooling towers; and (iv) implementing proactive and predictive maintenance plans to minimize equipment downtime and extend equipment life. Ultimately, a balance must be struck between commercial considerations, energy-optimized operation, and practical design.

6.5 Conclusions

In this chapter, the comparison of the DCS and CCS is made. Following the comparison, the eight major benefits of DCS systems are discussed in detail, aiming to solidity the understanding of DCS's potential for both technical and commercial advantages over CCS. Next, a comprehensive overview of the development process of different generations of refrigerants is provided. This helps readers grasp the advantages, disadvantages, environmental impacts, and global environmental protection efforts. Finally, the design and equipment selection of DCS system components should incorporate practical design and operational experiences from the DCS industry combined with insights from consulting, while also considering financial, and regulatory perspectives. The authors put the best effort to ensure the accuracy and reliability of the information based on the authors' industry and academic experience. However the authors provide no warranties of any kind, express or implied, about the completeness, accuracy, and reliability as most of the chiller plant design and equipment selections are heavily dependent on resource availability, technical consideration, and commercial considerations. (Note: Any reliance the reader places on such information is therefore strictly at the reader's own risk.)

References

1. Zaw, K., Kwik, Z.Z.S., Chang, W.Q., Islam, M.D., & Poh, T.K. (2023). A techno-commercial decision making framework for optimal district cooling system design in tropical regions. *Applied Thermal Engineering*, 220. https://doi.org/10.1016/j.applthermaleng.2022.119668

2. Werner, S. (2017). International review of district heating and cooling. *Energy*, 137, 617–631.
3. Dincer, I., & Zamfirescu, C. (2011). *Sustainable energy systems and applications*. Springer International Publishing.
4. Asian Development Bank. (2017). District cooling in the People's Republic of China: Status and development potential, creative commons attribution, Mandaluyong City, Phillippines. https://doi.org/10.22617/RPT168582-2
5. https://www.keppeldhcs.com/news-item.aspx?aid=2847&title=keppel-dhcs-recognised-as-solar-pioneer-2847
6. Jensen, M.K.. Marketing the case for district cooling. Dangoss A/S, Heating Segment, Nordborgvej 81, 6430 Nordborg, Denmark, 2016.
7. https://www.emsd.gov.hk/energyland/en/building/district_cooling_sys/dcs_benefits.html
8. Freon: A Collaborative Effort Among Three Companies (thoughtco.com)
9. http://refrigeranthq.com/dupont/
10. http://climate.nasa.gov/vital-signs/global-temperature/
11. http://www.cru.uea.ac.uk/
12. Dobson unit - Wikipedia
13. NASA SVS | Ozone Minimum Concentrations, 1979-2023
14. Montreal Protocol - Wikipedia
15. https://www.theozonehole.org/monthreal.htm
16. https://www.trane.com/content/dam/Trane/Commercial/north-america/about-us/newsroom/events/2023-engineer-summit/2023-EngSummit-Refrigerants-Update-FINAL.pdf
17. Green Mark 2021 | Building and Construction Authority (BCA)
18. United States Green Building Council. (2015). LEED v4. reference guide for building design + construction. [Online]. http://www.usgbc.org/guide/bdc
19. United States Green Building Council. (2010, August). District energy systems guidance v2.2 and v2009 (Guidance v2.0). Treatment of District or Campus Thermal Energy in LEED V2 and LEED 2009 – Design & Construction.
20. ASHRAE, IES, and ANSI. (2013). ANSI ASHRAE IES 90.1- 2013. Energy standard for buildings except low-rise residential buildings (I-P Edition)
21. Hartman, T. (2001). All variable speed centrifugal chiller plants. *ASHRAE Journal*, 43, 43–53.
22. AHRI, and ANSI. (2012). AHRI Standard 551/591 (SI) – 2011/ADD 3 Edition, Standard for performance rating of water-chilling and heat pump water-heating using the vapor compression cycle.
23. Hanlon, P. C. (2001). *Compressor handbook*. McGraw Hill.
24 Spring Singapore. (2014). Singapore standard 530:2014, Code of practice for energy.

7

Future Potential DCS Application Pathways and Conclusions

7.1 Introduction

District cooling systems (DCS) have emerged as a pivotal solution to meet the growing cooling demands in urban areas while addressing global climate change challenges. As cities expand and the need for cooling intensifies, the environmental impact of traditional cooling systems, reliant on fossil fuels, has come under increasing scrutiny. Carbon abatement, defined as the process of reducing or eliminating carbon dioxide (CO_2) emissions, plays a critical role in transitioning DCS toward sustainability. By focusing on both operational and embodied carbon emissions, DCS can significantly reduce their contribution to greenhouse gas (GHG) emissions. This chapter delves into the various strategies and technologies that support carbon abatement in DCS, examining their current status and future potential. From the integration of renewable energy sources such as solar, geothermal, and wind energy, to waste heat recovery techniques and the implementation of smart grid technologies, the discussion highlights the transformative potential of these innovations in reducing the carbon intensity of cooling systems. Furthermore, this chapter outlines a roadmap for achieving carbon-neutral DCS, emphasizing the importance of a comprehensive approach that includes technological advancements, policy support, and collaborative efforts.

7.1.1 Definition of Carbon Abatement

Carbon abatement refers to the process of reducing or eliminating carbon dioxide (CO_2) emissions and other GHGs released into the atmosphere. This reduction can be achieved through a variety of strategies, including improving energy efficiency, adopting renewable energy sources, implementing carbon capture and storage (CCS), and promoting sustainable practices in industries and infrastructure [1, 2]. In the context of district cooling and heating systems, carbon abatement involves minimizing the carbon footprint by

DOI: 10.1201/9781032670164-7

reducing emissions generated during energy production and distribution, as well as considering both operational and embodied carbon throughout the life cycle of the systems [3]. The goal of carbon abatement is to mitigate the environmental impact of human activities and contribute to global efforts to combat climate change.

7.1.2 Overview of Carbon Footprint of District Cooling Systems Globally

Despite the fact that DCS have many advantages over the conventional distributed cooling systems, DCS still contribute to global carbon emissions, particularly in regions where the electricity grid is heavily reliant on fossil fuels. The carbon footprint of DCS is influenced by several factors, including the energy source used to power the cooling plants, the efficiency of the distribution network, and the overall integration of renewable energy technologies into the system.

Globally, the carbon footprint of DCS is highly variable, depending on the energy mix and technological advancements within different regions. For instance, in countries with cleaner electricity grids that rely more on renewable energy sources such as hydroelectricity, wind, or solar, the carbon emissions from DCS are significantly lower. By contrast, in regions like the Middle East, where district cooling is widely deployed but where energy is predominantly sourced from natural gas and other fossil fuels, the carbon footprint remains high. Studies have shown that DCS can reduce energy consumption by 30%–50% compared to traditional cooling methods, but without a corresponding shift to low-carbon energy sources, the potential carbon reduction benefits remain limited [4].

The carbon footprint of DCS is also affected by the operational efficiency of the cooling plants and the design of the distribution network. Systems that incorporate advanced technologies such as thermal energy storage (TES), smart grid integration, and energy-efficient chillers can further reduce emissions. Additionally, the growing trend of integrating renewable energy sources like solar thermal energy and geothermal cooling into district cooling networks holds significant promise for reducing the overall carbon impact of these systems. However, despite technological advancements, many DCS around the world are still powered by conventional energy sources, resulting in a significant carbon footprint [5]. To fully realize the carbon reduction potential of district cooling, the integration of renewable energy and low-carbon technologies must be prioritized globally.

7.1.3 Importance of Carbon Abatement in the Context of District Cooling Systems

Carbon abatement plays a critical role in the evolution and sustainability of DCS. As urbanization accelerates and the demand for cooling grows, the environmental impact of traditional cooling methods has become increasingly concerning [1]. DCS offer a more efficient and centralized solution, but they still contribute to carbon emissions, primarily through the energy required to generate and distribute cooling. By focusing on carbon abatement, DCS can significantly reduce their GHG emissions, aligning with global efforts to mitigate climate change. This is particularly important as many cities and countries set ambitious targets for reducing carbon footprints and transitioning to net-zero carbon economies.

Implementing carbon abatement strategies in DCS can involve several approaches, such as integrating renewable energy sources, improving system efficiency, and utilizing waste heat or TES [6]. These methods not only reduce operational carbon emissions but also lower overall energy consumption, which is crucial for sustainable urban development. Moreover, addressing embodied carbon—the emissions associated with the construction, materials, and lifecycle of the system—further enhances the environmental benefits of district cooling. Ultimately, carbon abatement in DCS is not only a step toward a more sustainable future but also a necessary measure to ensure that cities can grow and thrive while minimizing their environmental impact.

7.2 Technologies Supporting Carbon Abatement in District Cooling

7.2.1 Integration of Renewable Energy Sources

The integration of renewable energy sources into DCS represents a critical strategy for achieving carbon abatement by directly reducing the reliance on fossil fuels and lowering associated GHG emissions. District cooling, when powered by conventional electricity sources, can contribute significantly to carbon emissions, especially in regions where the electricity grid is dominated by coal or natural gas. By integrating clean and renewable energy sources, such as solar and geothermal, DCS can reduce their carbon footprint and contribute to broader global efforts to mitigate climate change..

7.2.1.1 Integration of Solar Cooling Systems Into DCS

Solar cooling is an ideal solution for reducing carbon emissions in regions with high solar potential. Solar thermal energy, captured through collectors or concentrated solar power (CSP), can be used to power absorption chillers that produce cooling without the need for electricity generated from fossil fuels. This can lead to substantial reductions in CO_2 emissions. For example, in the **Masdar City Solar Cooling Plant** in Abu Dhabi, the use of solar thermal energy in district cooling has significantly reduced the city's reliance on conventional power plants, which are predominantly fueled by natural gas [7]. By offsetting this need, the city reduces its operational carbon emissions and overall carbon footprint. The system also incorporates TES, which allows excess solar energy to be stored and used during periods of peak cooling demand or when sunlight is unavailable. This not only optimizes energy use but also reduces the emissions associated with generating electricity during peak hours, when fossil fuel power plants are typically employed. The result is a DCS that aligns with the city's broader sustainability goals, contributing to long-term carbon abatement.

7.2.1.2 Integration of Geothermal Cooling Systems into DCS

Geothermal cooling is another renewable energy source that can drastically lower the carbon emissions of DCS. Geothermal systems tap into the Earth's stable underground temperatures to provide year-round cooling without the need for energy-intensive refrigeration processes. By replacing conventional cooling technologies with geothermal energy, DCS can significantly reduce their reliance on electricity generated from fossil fuels, thereby lowering carbon emissions [8]. In **Reykjavik, Iceland**, the geothermal DCS reduce the city's carbon footprint by using naturally cool geothermal water for cooling, instead of relying on electricity to power mechanical chillers [9–10]. Similarly, **Cornell University's Lake Source Cooling Project** in New York uses deep lake water as a cooling source, reducing the university's energy consumption by up to 90% compared to conventional systems [11]. By drawing on renewable cooling sources, these systems dramatically cut energy use and the associated carbon emissions, contributing to a more sustainable cooling solution.

7.2.1.3 Integration of Biomass and Wind Energy into DCS

The integration of **biomass** and **wind energy** into DCS, while less common than solar and geothermal, also plays a significant role in carbon abatement. Biomass energy, generated from organic materials such as wood chips, agricultural waste, or dedicated energy crops, can be used to power district cooling plants. For instance, the **Malmö Västra Hamnen District Cooling System** in Sweden uses biomass as part of a combined heat and power (CHP) process that provides energy for both heating and cooling [12]. Biomass is considered carbon-neutral because the CO_2 released during combustion is offset by the carbon absorbed during the growth of the biomass feedstock. This makes it a sustainable energy source that reduces overall emissions. Additionally, wind energy is increasingly being used to supply electricity for DCS. In **Copenhagen, Denmark**, the city's DCS are powered in part by electricity from offshore wind farms. By shifting to a cleaner energy grid that includes wind power, Copenhagen's district cooling network reduces its reliance on fossil fuels, helping to significantly lower its carbon emissions [13]. Wind energy provides a renewable, non-polluting power source that can dramatically reduce the carbon intensity of district cooling operations.

The integration of renewable energy into DCS contributes directly to **carbon abatement** by replacing or reducing the need for fossil fuel-generated electricity. Each of these renewable energy sources offers unique benefits for reducing the carbon emissions of cooling systems. Solar cooling offsets the need for electricity during peak demand periods, geothermal systems provide low-carbon cooling year-round, and biomass and wind energy offer carbon-neutral or low-carbon alternatives to traditional energy sources. As these technologies continue to mature and scale, they will play an increasingly important role in reducing the environmental impact of DCS and helping cities meet their carbon reduction targets. The combination of renewable energy integration, increased efficiency, and lower operational emissions makes DCS a powerful tool in the fight against climate change, contributing to both immediate carbon reductions and long-term sustainability goals.

7.2.2 Waste Heat Recovery for District Cooling

Waste heat recovery is an increasingly important technology for reducing the carbon footprint of DCS. By capturing and repurposing heat generated from industrial processes, power generation, or even building operations, DCS can use this excess thermal energy to provide cooling without the need for additional energy input from fossil fuel sources. This not only improves the energy efficiency of the cooling system but also significantly reduces CO_2 emissions [14]. Waste heat recovery offers a valuable opportunity for carbon abatement in the cooling sector, contributing to both operational savings and environmental sustainability.

One of the most common waste heat recovery techniques used in district cooling is CHP, also known as cogeneration. In a CHP system, electricity and heat are produced simultaneously from the same energy source. The heat that is typically wasted in conventional power generation is captured and used for cooling via absorption chillers. These chillers utilize waste heat to drive the cooling process, turning what would otherwise be discarded energy into a useful output. For example, Empower, a district cooling company in Dubai, has integrated waste heat from CHP plants to drive absorption chillers in their district cooling networks. By capturing the waste heat from local power plants, Empower is able to generate cooling without relying entirely on grid electricity, leading to a significant reduction in the system's carbon emissions. This system not only boosts energy efficiency but also helps meet the cooling demands of regions with extreme climates in a more sustainable way.

Another waste heat recovery method involves industrial process heat, where excess heat from industrial operations such as steel manufacturing, refineries, or data centers is captured and used in district cooling. In Helsinki, Finland, the Katri Vala Heating and Cooling Plant captures waste heat from data centers and sewage treatment plants and repurposes it for district heating and cooling [15]. The heat is absorbed and converted into cooling energy using heat pumps, which then provide chilled water to the district cooling network. This process significantly reduces the need for conventional energy sources for cooling, thus mitigating CO_2 emissions. By using waste heat from these industrial processes, the system lowers both operational costs and carbon emissions, enhancing the overall sustainability of the district cooling network.

Waste heat can also be recovered from commercial buildings and urban infrastructure. Large commercial buildings such as shopping malls, hotels, and hospitals often generate significant amounts of waste heat from cooling and ventilation systems, which can be captured and reused. In Paris, France, the Climespace District Cooling System recovers waste heat from commercial buildings and uses it to provide cooling across the city's district network. The heat recovered from heating ventilation and air conditioning (HVAC) systems is channeled back into the district energy loop, where it is used in absorption chillers to generate chilled water for cooling other buildings. This method not only reduces the energy demand of individual buildings but also decreases the overall carbon emissions of the DCS by recycling energy that would otherwise be lost.

The implementation of waste heat recovery in district cooling requires the integration of advanced technologies such as absorption chillers, heat exchangers, and TES systems. Absorption chillers play a key role in converting waste heat into cooling by using heat to drive a refrigerant cycle rather than electricity. Heat exchangers allow the efficient transfer of thermal energy from the waste heat source to the cooling system. In addition, TES can be used to store excess cooling produced from waste heat during periods of low demand for use during peak periods. This combination of technologies ensures that the waste heat is captured and utilized in the most efficient manner possible, maximizing both energy savings and carbon reductions.

The carbon abatement potential of waste heat recovery in DCS is significant. By reducing the reliance on electricity generated from fossil fuels, waste heat recovery can lead to substantial reductions in CO_2 emissions. For instance, the Copenhagen District Cooling System in Denmark, which integrates waste heat from local power plants and industrial sources, has reduced its carbon emissions by more than 70% compared to conventional cooling systems. Waste heat recovery helps reduce the overall energy intensity of DCS, making them more sustainable and contributing to national and global carbon reduction targets.

This method also complements the increasing focus on circular energy systems, where waste energy is repurposed and recycled, thereby improving resource efficiency and reducing environmental impact.

7.2.3 Smart Grid Technologies and Real-Time Optimization for Energy Efficiency

The integration of smart grid technologies and real-time optimization into DCS represents a significant advancement in enhancing energy efficiency and reducing carbon emissions. Smart grids use digital technology to monitor, control, and optimize energy production and consumption in real time, ensuring that energy is distributed and used as efficiently as possible. In the context of district cooling, smart grid technologies enable dynamic load balancing, predictive demand management, and seamless integration with renewable energy sources, all of which contribute to significant reductions in energy consumption and carbon emissions.

Smart grid technologies enhance energy efficiency in DCS by providing real-time data on energy use, which allows for more precise adjustments in cooling output to match fluctuating demand. For instance, sensors and smart meters embedded in the cooling network can continuously monitor the temperature, humidity, and cooling load of individual buildings or zones. This data is then analyzed to optimize the operation of chillers, pumps, and distribution networks, ensuring that cooling is provided only when and where it is needed. The ability to adjust cooling in real time prevents energy waste and reduces the need for overproduction, which is a common inefficiency in traditional systems.

One of the key tools for energy optimization in smart grids is **demand response**. Demand response programs allow DCS to adjust their cooling load in response to real-time changes in electricity prices or grid conditions. For example, during periods of high electricity demand, a smart grid can temporarily reduce cooling output to lower overall energy consumption, thereby reducing the strain on the grid. This not only cuts energy use during peak hours but also reduces the need for utilities to activate additional, often carbon-intensive, power plants. In **New York City, Con Edison's district cooling system** uses smart grid technologies and demand response to dynamically adjust cooling loads based on real-time grid conditions. This system has significantly improved energy efficiency, especially during peak periods, while also reducing carbon emissions.

The implementation of real-time optimization in DCS requires the integration of **advanced control systems**, **smart sensors**, **predictive analytics**, and **machine learning algorithms**. First, a network of smart sensors is installed throughout the DCS, including key points such as chillers, pumps, and thermal storage units. These sensors continuously collect data on system performance, energy use, and environmental conditions, which are then fed into a central control platform.

Predictive analytics and **machine learning algorithms** are used to process this data, predict future cooling demand, and identify areas where energy savings can be made. By anticipating demand peaks, the system can pre-emptively adjust cooling output or activate TES systems to meet demand without increasing energy consumption. For example, in **Singapore's Marina Bay District Cooling System**, real-time optimization algorithms are used to manage the network's chillers and thermal storage, allowing the system to respond instantly to changes in cooling demand. This has resulted in a significant reduction in energy consumption and operational costs, as the system can operate more efficiently and avoid unnecessary energy use.

Smart grids can facilitate the seamless integration of solar, wind, or geothermal energy into the cooling system by adjusting the energy mix based on availability. When renewable energy generation is high, such as during sunny or windy days, the cooling system can prioritize the use of renewable energy, reducing reliance on fossil fuels. When renewable energy availability is low, the system can switch to stored energy or reduce cooling output during non-peak times to maintain efficiency. This flexibility is key to reducing carbon emissions, as it allows DCS to maximize their use of clean energy sources.

The deployment of smart grid technologies and real-time optimization has a direct and substantial impact on carbon emissions. By reducing energy consumption and enhancing system efficiency, these technologies cut the overall demand for electricity, which is often generated from fossil fuels. Additionally, demand response programs lower the need for carbon-intensive peaking power plants, which are typically activated during periods of high demand. This not only reduces the carbon intensity of the electricity grid but also minimizes the environmental impact of DCS.

The **Copenhagen District Cooling System** is an example of how smart grid technology can be used to achieve carbon abatement [17]. The system integrates real-time optimization and demand response to adjust cooling loads based on electricity availability and grid conditions. Additionally, it utilizes wind energy and TES to further reduce its reliance on fossil fuel-based electricity. Through the use of smart grid technologies, the system has achieved a 30% reduction in energy consumption and has significantly lowered carbon emissions. Moreover, by prioritizing the use of renewable energy sources, the system has contributed to the city's broader goal of becoming carbon-neutral by 2025.

In **Stockholm, Sweden**, the DCS also integrates smart grid technologies to improve energy efficiency and reduce carbon emissions. The system uses real-time data to optimize the operation of heat pumps and TES, allowing it to reduce energy consumption during peak periods. By integrating renewable energy and improving operational efficiency, the Stockholm system has reduced its CO_2 emissions by more than 50% compared to traditional cooling methods.

7.2.4 Enhancement in Chiller Technology and Its Impact on Carbon Emissions

Advancements in chiller technology have significantly enhanced the efficiency of DCS, leading to substantial reductions in energy consumption and carbon emissions. Traditional chillers, often powered by electricity generated from fossil fuels, have historically been a major contributor to the carbon footprint of cooling systems. However, recent innovations in chiller design, particularly in the areas of variable-speed drives (VSDs), magnetic bearing compressors, and absorption chillers, have dramatically improved the energy efficiency of chillers. These improvements have reduced the electricity demand of DCS, thus lowering their reliance on carbon-intensive energy sources and directly contributing to carbon abatement.

One of the key enhancements in chiller technology is the introduction of VSDs, which allow chillers to adjust their output based on real-time cooling demand. Unlike conventional chillers that operate at a constant speed, regardless of demand, VSD-enabled chillers can modulate their energy use in response to fluctuating loads. This not only reduces energy waste but also optimizes system performance, leading to significant energy savings. For example, a study of chillers with VSDs in Singapore's DCS revealed up to 30% energy savings compared to traditional chillers, translating into a substantial reduction in carbon emissions. Similarly, magnetic bearing compressors have improved the operational efficiency of chillers by reducing mechanical friction, which in turn lowers the energy required to maintain cooling output.

In addition to mechanical advancements, absorption chillers represent a major leap in reducing carbon emissions. Absorption chillers use heat (often from waste heat or renewable energy sources like solar thermal energy) instead of electricity to drive the refrigeration cycle. This makes them ideal for DCS, aiming to reduce their reliance on fossil fuel-powered electricity. By leveraging waste heat or renewable energy, absorption chillers can deliver cooling while emitting far less CO_2 compared to electric-powered chillers. For instance, in the Masdar City Solar Cooling Plant in the UAE, absorption chillers powered by solar energy have been successfully integrated into the district cooling network, significantly reducing both energy use and carbon emissions.

7.3 Future Developments and Innovations

7.3.1 Emerging Technologies That Could Drive Further Carbon Reductions (e.g. AI for System Optimization, Advanced Heat Pumps)

As global demand for DCS grows in response to urbanization and climate change, emerging technologies are playing a pivotal role in driving further carbon reductions in these systems. The need for low-carbon cooling solutions is more pressing than ever, given the significant energy consumption and associated carbon emissions of traditional cooling methods. Innovative technologies such as advanced energy storage, AI-driven optimization, high-temperature cooling, and advanced heat pump systems are emerging as critical tools in reducing the carbon footprint of DCS.

TES systems are becoming an increasingly important component in district cooling infrastructure. TES systems allow excess cooling energy to be stored during periods of low demand and used during peak hours, thus flattening demand peaks and reducing the need for carbon-intensive energy during high-demand periods. For instance, TES can store chilled water or ice produced during off-peak hours, when renewable energy sources like wind or solar are more available and then discharge this stored cooling energy during peak hours when demand is highest. This not only enhances the operational efficiency of DCS but also minimizes the reliance on fossil fuel-powered electricity during peak periods. In Toronto's Deep Lake Water Cooling (DLWC) system, for example, TES plays a critical role in maintaining efficient energy use, enabling the system to deliver cooling with significantly lower carbon emissions compared to conventional systems.

AI and machine learning for real-time optimization are also emerging as transformative technologies for carbon reduction in DCS. These technologies leverage advanced algorithms and data analytics to continuously monitor, optimize, and adjust the performance of district cooling networks. By using predictive analytics, AI-driven systems can anticipate cooling demand patterns based on weather forecasts, historical usage data, and building occupancy, allowing the system to make real-time adjustments to improve energy efficiency. This level of optimization ensures that cooling systems operate at peak efficiency with minimal energy waste. For example, in Singapore's Marina Bay District Cooling system, real-time optimization using AI has resulted in energy savings of up to 20%, significantly reducing the carbon footprint of the system. The integration of AI technologies allows for smarter, more adaptive cooling solutions that are aligned with the broader goals of energy efficiency and carbon abatement.

High-temperature district cooling is another emerging technology that holds potential for driving down carbon emissions in cooling systems. Traditional district cooling operates at relatively low temperatures, which can result in significant energy losses during distribution. High-temperature cooling systems, however, are designed to operate at higher temperatures, which reduces thermal losses and improves the overall efficiency of the cooling process. By reducing the temperature difference between the cooling medium and the external environment, high-temperature cooling systems can significantly reduce the energy required for cooling. Additionally, these systems are easier to integrate with waste heat recovery and renewable energy sources, further enhancing their sustainability profile. For example, the EcoGrid 2.0 project in Sweden is exploring high-temperature district cooling combined with renewable energy integration, which has shown promising results in reducing energy consumption and carbon emissions [16].

Advanced heat pump technology is also playing a critical role in reducing carbon emissions in DCS. Heat pumps, which transfer heat from one place to another rather than generating it, are highly efficient and can be powered by renewable energy sources. New advancements in heat pump technology, including hybrid systems that combine electric and absorption heat pumps, are improving the performance and efficiency of DCS. These advanced systems can simultaneously provide both heating and cooling, using minimal energy to meet the demand for both services. The use of low-carbon heat pumps in combination with renewable energy, as seen in Stockholm Exergi's DCS, has already led to significant carbon reductions and is setting new standards for the future of low-carbon district cooling.

Finally, decentralized and modular cooling systems are gaining attention as an emerging solution that could enhance the flexibility and efficiency of district cooling networks. Modular cooling systems allow for the deployment of smaller, localized cooling units that can operate independently or as part of a larger network. This decentralized approach reduces the need for long-distance energy transport, which is often associated with thermal losses and inefficiencies. Additionally, modular systems can be more easily integrated with renewable energy sources, allowing for a more flexible and scalable solution that can adapt to the specific needs of different areas within a city. Decentralized DCS are currently being tested in cities such as Copenhagen, where they are expected to reduce carbon emissions by minimizing distribution losses and optimizing energy use on a local scale.

7.3.2 Prospects for Integrating District Cooling with Broader Urban Sustainability Initiatives

The integration of DCS with broader urban sustainability initiatives offers a significant opportunity to enhance the efficiency, resilience, and environmental performance of cities. As global cities increasingly commit to ambitious sustainability goals, district cooling has the potential to be a key enabler of low-carbon urban development. District cooling, by its nature, supports the goals of energy efficiency, renewable energy integration, and the reduction of GHG emissions. When aligned with larger urban sustainability strategies such as green building standards, smart cities initiatives, and climate resilience planning, district cooling can play a transformative role in creating more sustainable, livable urban environments [18].

One of the primary ways district cooling aligns with urban sustainability initiatives is through its ability to optimize energy use in dense, urban settings. DCS replaces inefficient, decentralized cooling systems with a centralized infrastructure that can distribute cooling more efficiently across multiple buildings, reducing overall energy consumption. This aligns with the broader goal of energy efficiency in cities, which is often a key component of sustainability frameworks such as LEED (Leadership in Energy and Environmental Design) and BREEAM (Building Research Establishment Environmental Assessment Method). Cities such as Dubai, which has mandated the use of district cooling in new developments, are integrating this technology as part of their green building initiatives to reduce the urban heat island effect, minimize electricity demand, and achieve long-term carbon reduction targets. By aggregating cooling loads, DCS can reduce peak electricity demand, making cities more energy-efficient and reducing the strain on local energy grids.

Furthermore, the integration of district cooling with renewable energy initiatives is crucial for advancing urban sustainability goals. As cities transition toward clean energy, DCS can be designed to incorporate renewable energy sources such as solar thermal, geothermal, and biomass. For example, in Masdar City, Abu Dhabi, the DCS is powered by solar energy, reducing the city's reliance on conventional power sources and contributing to its goal of being one of the most sustainable urban developments in the world [7]. The coupling of district cooling with renewable energy sources not only lowers operational carbon emissions but also creates opportunities for cities to meet their renewable energy targets. As cities increasingly adopt renewable energy portfolios, DCS will play an essential role in maximizing the use of clean energy in urban cooling.

Another prospect for integrating district cooling into urban sustainability initiatives is its role in supporting climate resilience. As cities become more vulnerable to the effects of climate change, including rising temperatures and increased energy demand for cooling, district cooling can provide a resilient infrastructure for managing thermal loads. By leveraging technologies such as TES and smart grid systems, DCS can enhance a city's ability to maintain reliable and energy-efficient cooling during heat waves or power shortages. This makes district cooling an important tool in urban climate adaptation strategies, as it can mitigate the risks associated with extreme heat events, which are expected to become more frequent and severe as global temperatures rise. In Copenhagen, for instance, the city's district cooling network is designed to help mitigate the effects of climate change by providing energy-efficient cooling to critical urban infrastructure, such as hospitals and data centers, while reducing the city's overall carbon footprint.

Moreover, the potential for integrating DCS with smart cities initiatives offers a forward-looking opportunity to enhance urban sustainability. Smart cities aim to use digital technology to optimize urban systems, including energy, water, and transportation, to improve quality of life while reducing environmental impact. District cooling, when integrated with smart grid technologies and Internet of Things (IoT) solutions, can become a key component of a city's smart infrastructure. Smart grid systems allow district cooling networks to dynamically adjust cooling loads in response to real-time data, optimizing energy use based on factors such as weather conditions, building occupancy, and electricity prices. This level of optimization not only improves energy efficiency but also ensures that DCS can operate seamlessly alongside other urban systems, contributing to the overall sustainability and resilience of the city.

7.3.3 Roadmap for Achieving Carbon-Neutral District Cooling Systems

Achieving carbon-neutral DCS requires a systematic and multi-faceted approach that integrates technological innovation, renewable energy adoption, policy support, and stakeholder engagement. The pathway toward carbon neutrality involves reducing operational and embodied carbon emissions while ensuring that cooling demands in urban areas are met in a sustainable and resilient manner. A well-structured roadmap provides the framework for cities, energy providers, and policymakers to collaborate on creating DCS that align with global carbon reduction targets and sustainability goals.

The first step in this roadmap is maximizing energy efficiency through advanced technologies. DCS must implement the latest innovations in energy-efficient chillers, smart grid integration, and real-time optimization technologies. High-efficiency chillers, particularly those with VSDs and magnetic bearing compressors, can significantly reduce the energy required to cool large urban areas. Integrating smart grid technologies with IoT devices allows for real-time monitoring and dynamic load balancing, optimizing energy use based on demand fluctuations. AI and machine learning algorithms can predict cooling needs [19], enabling DCS to adjust operations preemptively, thereby reducing energy waste and carbon emissions. By prioritizing energy efficiency at every level of the cooling network, cities can reduce the operational carbon footprint of DCS and create a foundation for future decarbonization.

The second critical phase involves the integration of renewable energy sources into DCS. To achieve carbon neutrality, district cooling networks must transition away from fossil fuels and incorporate a larger share of renewable energy in their operations. Solar thermal energy, geothermal cooling, wind power, and biomass are promising renewable sources that can power DCS with minimal carbon emissions. For instance, solar cooling can be utilized in regions with high solar irradiance, where solar thermal collectors can drive absorption chillers to provide cooling, significantly reducing reliance on grid electricity. In colder climates, geothermal energy offers a reliable, year-round source of cooling. Furthermore, waste heat recovery from industrial processes or power plants can be integrated into district cooling networks, turning excess heat into useful cooling energy. By coupling renewable energy with waste heat recovery and other low-carbon technologies, DCS can drastically reduce their reliance on carbon-intensive power sources and move closer to carbon neutrality.

The roadmap also calls for enhanced TES capabilities. TES plays a vital role in bridging the gap between energy supply and demand, allowing DCS to store excess energy during periods of low demand (or when renewable energy is abundant) and use it during peak demand periods. By reducing the need to generate additional cooling capacity during peak hours, TES improves the overall efficiency of district cooling networks while helping to balance the energy grid. Cities like Toronto and Singapore have successfully integrated TES into their district cooling networks, significantly lowering peak demand energy use and enabling the more effective integration of renewable energy sources. Expanding the use of TES across DCS will be a key factor in achieving carbon-neutral operations.

Another important element in the roadmap is the adoption of circular energy practices. Circular energy systems emphasize the reuse, recycling, and sharing of energy resources to minimize waste and carbon emissions. In the context of district cooling, this involves utilizing waste heat from nearby industrial or commercial processes, repurposing cooling energy, and integrating with other urban energy systems such as district heating. For example, Copenhagen's district energy system integrates both heating and cooling, with waste heat from cooling processes being reused to provide heating to other parts of the city. This type of synergy between district cooling and other urban energy networks can significantly improve energy efficiency, reduce carbon emissions, and lower operational costs.

The final, yet critical, component of the roadmap involves regulatory and policy support. Governments and local authorities are crucial in facilitating the transition to carbon-neutral DCS. Policies that incentivize the adoption of renewable energy, impose carbon pricing, and support energy-efficient building standards are essential for driving the widespread implementation of carbon-neutral DCS. For example, urban planning regulations could mandate the use of district cooling in new large-scale developments, while offering subsidies or tax credits to developers that incorporate renewable energy or energy-efficient technologies into their cooling systems. Additionally, carbon-neutral district cooling requires a collaborative approach, involving energy providers, city planners, developers, and residents. Public–private partnerships (PPPs) can play a pivotal role in securing investment for large-scale infrastructure projects and ensuring that district cooling networks are designed and operated in a way that aligns with both public and private sector sustainability goals.

7.4 Conclusion

7.4.1 Summary of Current Achievements in Carbon Abatement for District Cooling

In conclusion, DCS present a powerful solution for reducing the carbon footprint of urban cooling, with numerous technological advancements and innovative practices driving significant progress in carbon abatement. The integration of waste heat recovery techniques such as CHP, industrial process heat capture, and urban heat recovery has already been demonstrated in cities worldwide, showing how otherwise wasted thermal energy can be repurposed to reduce operational emissions. Meanwhile, smart grid technologies and real-time optimization are revolutionizing energy efficiency in DCS by using data analytics and demand response to optimize performance, reduce energy consumption, and enable the integration of renewable energy sources. Cities such as New York, Copenhagen, and Stockholm showcase the potential of these innovations to dramatically lower energy use and emissions. In tandem with enhancements in chiller technology—such as high-efficiency chillers and renewable energy-driven absorption chillers—DCS are becoming more efficient and less carbon-intensive, advancing the global agenda for carbon abatement and climate change mitigation.

Looking ahead, emerging technologies such as TES, AI-driven optimization, high-temperature cooling, and decentralized cooling systems hold immense potential to further revolutionize district cooling. These innovations will not only reduce energy consumption but also align with broader urban sustainability initiatives such as smart city development, climate resilience planning, and renewable energy integration. The future roadmap for achieving carbon-neutral district cooling involves a comprehensive strategy, focusing on maximizing energy efficiency, integrating low-carbon energy sources, adopting circular energy practices, and fostering supportive policy frameworks. As cities around the world commit to ambitious climate goals, DCS will play an increasingly critical role in their ability to meet these targets, creating more sustainable, resilient urban environments while reducing their overall carbon footprints.

7.4.2 Future Outlook for Carbon Reductions in District Cooling Systems

The future outlook for carbon reductions in DCS is highly promising, driven by advancements in technology, increased adoption of renewable energy sources, and evolving regulatory frameworks that prioritize sustainability. As urbanization continues and global temperatures rise, the demand for cooling is expected to grow exponentially, making it imperative for DCS to evolve in ways that significantly reduce their carbon footprints. The development and implementation of innovative technologies, such as advanced chiller systems, TES, and AI-driven optimization, will be central to achieving this goal. These technologies will not only enhance the energy efficiency of DCS but also enable the integration of low-carbon and renewable energy sources such as solar thermal, geothermal, and wind energy. As these technologies mature and scale, their contribution to decarbonizing district cooling networks will become increasingly significant.

A critical component of future carbon reductions in district cooling will be the wider adoption of circular energy systems and waste heat recovery techniques. By capturing and repurposing waste heat from industrial processes, power plants, and urban infrastructure, DCS can further decrease their reliance on fossil fuels and enhance their overall energy efficiency. Additionally, the integration of smart grid technologies will play a transformative role in reducing carbon emissions. Real-time optimization, predictive demand management, and demand response mechanisms will allow DCS to operate more dynamically, optimizing energy use based on fluctuating demand and the availability of renewable energy. Governments and cities are also expected to play an essential role by adopting supportive policies, including carbon pricing, incentives for energy-efficient technologies, and mandates for renewable energy integration in new developments. These collective efforts will drive substantial carbon reductions in district cooling, positioning it as a cornerstone in the global transition to sustainable, low-carbon urban energy systems.

References

[1] Bandyopadhyay, B., & Banerjee, M. (2022). Decarbonization of cooling of buildings. *Solar Compass*, 2, 100025. https://doi.org/10.1016/j.solcom.2022.100025

[2] E.K. Tetteh, M.O. Amankwa, C. Yeboah, & M.O. Amankwa (2021). Emerging carbon abatement technologies to mitigate energy-carbon footprint – A review. *Cleaner Materials*, 2, 100020. ISSN 2772-3976, https://doi.org/10.1016/j.clema.2021.100020

[3] Samir Idrissi Kaitouni, Fatime-Zohra Gargab, Ahmed Tabit, Mustapha Mabrouki, Nouzha Lamdouar, Abdelmajid Jamil, & Mohamed Ahachad (2024). A life cycle carbon dioxide equivalent emissions assessment of zero carbon building in hot semi-arid climate region: Case study. *Results in Engineering*, 23, 102589. ISSN 2590-1230. https://doi.org/10.1016/j.rineng.2024.102589

[4] Mohammed Aloshan, & Kareem Aldali (2024). Empirical study of facade retrofits for optimizing energy efficiency and cooling in school buildings in Saudi Arabia. *Energy Reports*, 12, 4105–4128. ISSN 2352-4847 https://doi.org/10.1016/j.egyr.2024.09.076

[5] Zhicheng Duan, Tie Wei, Pin Xie, Yilong Lu (2024). Co-benefits and influencing factors exploration of air pollution and carbon reduction in China: Based on marginal abatement costs. *Environmental Research*, 252(Part 1), 118742. ISSN 0013-9351. https://doi.org/10.1016/j.envres.2024.118742

[6] Ryan Janzen, Matthew Davis, & Amit Kumar (2020). Greenhouse gas emission abatement potential and associated costs of integrating renewable and low carbon energy technologies into the Canadian oil sands. *Journal of Cleaner Production*, 272, 122820. ISSN 0959-6526. https://doi.org/10.1016/j.jclepro.2020.122820

[7] Afif Akel Hasan, Adel Juaidi, Ramez Abdallah, Tareq Salameh, Osama Ayadi, Mustafa Jaradat, Raghad Emad Hammad, Pietro Elia Campana, Omar Afif Aqel (2022). A review of solar thermal cooling technologies in selected Middle East and North African countries. *Sustainable Energy Technologies and Assessments*, 54, 102871. ISSN 2213-1388, https://doi.org/10.1016/j.seta.2022.102871

[8] Vaishnavi Pandey, Anirbid Sircar, Namrata Bist, Kriti Yadav, & Dharmesh Morabiya Design of a space cooling plant from residual energy of ORC-GSHP assisted geothermal power plant at Dholera, Gujarat, India, *Applied Thermal Engineering*, 231, 2023, 120941. ISSN 1359-4311. https://doi.org/10.1016/j.applthermaleng.2023.120941

[9] Gudni Axelsson, Einar Gunnlaugsson, Thorgils Jónasson, & Magnús Ólafsson (2010). Low-temperature geothermal utilization in Iceland – Decades of experience. *Geothermics*, 39(4), 329–338. ISSN 0375-6505. https://doi.org/10.1016/j.geothermics.2010.09.002

[10] Alexandra Kjeld, Helga Johanna Bjarnadottir, & Ragnheiður Olafsdottir (2022). Life cycle assessment of the Theistareykir geothermal power plant in Iceland. *Geothermics*, 105, 102530. ISSN 0375-6505. https://doi.org/10.1016/j.geothermics.2022.102530

[11] Xueyu Tian, Yilun Zhou, Brianna Morris, & Fengqi You (2022). Sustainable design of Cornell University campus energy systems toward climate neutrality and 100% renewables. *Renewable and Sustainable Energy Reviews*, 161, 112383. ISSN 1364-0321, https://doi.org/10.1016/j.rser.2022.112383

[12] Janneke van der Leer, Alexandra Calvén, Wiktoria Glad, Paula Femenías, & Kerstin Sernhed (2023). Energy systems in sustainability-profiled districts in Sweden: A literature review and a socio-technical ecology approach for future research. *Energy Research & Social Science*, 101, 103118, ISSN 2214-6296. https://doi.org/10.1016/j.erss.2023.103118

[13] Yuhang Zhang, Mingzhe Liu, Zheng O'Neill, & Jin Wen (2024). Temperature control strategies for fifth generation district heating and cooling systems: A review and case study. *Applied Energy*, 376(Part A), 124156. ISSN 0306-2619. https://doi.org/10.1016/j.apenergy.2024.124156

[14] Myat, A., Thu, K., Kim, Y. D., Saha, B. B., & Ng, K. C. (2012). Entropy generation minimization: A practical approach for performance evaluation of temperature cascaded co-generation plants. *Energy*, 46(1), 493–521. ISSN 0360-5442. https://doi.org/10.1016/j.energy.2012.07.062

[15] Mikko Wahlroos, Matti Pärssinen, Samuli Rinne, Sanna Syri, Jukka Manner (2018). Future views on waste heat utilization – Case of data centers in Northern Europe. *Renewable and Sustainable Energy Reviews*, 82(Part 2), 1749–1764. ISSN 1364-0321. https://doi.org/10.1016/j.rser.2017.10.058

[16] Carsten Heinrich, Charalampos Ziras, Angeliki L.A. Syrri, & Henrik W. Bindner (2020). EcoGrid 2.0: A large-scale field trial of a local flexibility market. *Applied Energy*, 261, 114399. ISSN 0306-2619. https://doi.org/10.1016/j.apenergy.2019.114399

[17] Rex Martin, Simon Peter Aslak Kondrup Larsen (2024). "I never look at a temperature device, I just feel it": Practical knowledge, smart technologies, and heating and cooling practices between Denmark and Australia. *Energy Research Social Science*, 108, 103389. ISSN 2214-6296, https://doi.org/10.1016/j.erss.2023.103389

[18] Khosla, R., Miranda, N.D., Trotter, P.A. et al. (2021). Cooling for sustainable development. *Nat Sustain* **4**, 201–208. https://doi.org/10.1038/s41893-020-00627-w

[19] Myat, A., Kondath, N., Soh, Y. L., Hui, A. (2024). A hybrid model based on multivariate fast iterative filtering and long short-term memory for ultra-short-term cooling load prediction. *Energy and Buildings*, 307, 113977, ISSN 0378-7788, https://doi.org/10.1016/j.enbuild.2024.113977

Index

Pages in *italics* refer to figures and pages in **bold** refer to tables.

I

K

L

M

N

T

U

V

W